T0296072

LONDON MATHEMATICAL SOCIETY LECTURE NOTE SERIES

Managing Editor: Professor M. Reid, Mathematics Institute,
University of Warwick, Coventry CV4 7AL, United Kingdom

The titles below are available from booksellers, or from Cambridge University Press at
http://www.cambridge.org/mathematics

362   Differential tensor algebras and their module categories,   R. BAUTISTA, L. SALMERÓN & R. ZUAZUA
363   Foundations of computational mathematics, Hong Kong 2008,   F. CUCKER, A. PINKUS & M.J. TODD (eds)
364   Partial differential equations and fluid mechanics,   J.C. ROBINSON & J.L. RODRIGO (eds)
365   Surveys in combinatorics 2009,   S. HUCZYNSKA, J.D. MITCHELL & C.M. RONEY-DOUGAL (eds)
366   Highly oscillatory problems,   B. ENGQUIST, A. FOKAS, E. HAIRER & A. ISERLES (eds)
367   Random matrices: High dimensional phenomena,   G. BLOWER
368   Geometry of Riemann surfaces,   F.P. GARDINER, G. GONZÁLEZ-DIEZ & C. KOUROUNIOTIS (eds)
369   Epidemics and rumours in complex networks,   M. DRAIEF & L. MASSOULIÉ
370   Theory of p-adic distributions,   S. ALBEVERIO, A.YU. KHRENNIKOV & V.M. SHELKOVICH
371   Conformal fractals,   F. PRZYTYCKI & M. URBAŃSKI
372   Moonshine: The first quarter century and beyond,   J. LEPOWSKY, J. MCKAY & M.P. TUITE (eds)
373   Smoothness, regularity and complete intersection,   J. MAJADAS & A. G. RODICIO
374   Geometric analysis of hyperbolic differential equations: An introduction,   S. ALINHAC
375   Triangulated categories,   T. HOLM, P. JØRGENSEN & R. ROUQUIER (eds)
376   Permutation patterns,   S. LINTON, N. RUŠKUC & V. VATTER (eds)
377   An introduction to Galois cohomology and its applications,   G. BERHUY
378   Probability and mathematical genetics,   N. H. BINGHAM & C. M. GOLDIE (eds)
379   Finite and algorithmic model theory,   J. ESPARZA, C. MICHAUX & C. STEINHORN (eds)
380   Real and complex singularities,   M. MANOEL, M.C. ROMERO FUSTER & C.T.C WALL (eds)
381   Symmetries and integrability of difference equations,   D. LEVI, P. OLVER, Z. THOMOVA & P. WINTERNITZ (eds)
382   Forcing with random variables and proof complexity,   J. KRAJÍČEK
383   Motivic integration and its interactions with model theory and non-Archimedean geometry I,   R. CLUCKERS, J. NICAISE & J. SEBAG (eds)
384   Motivic integration and its interactions with model theory and non-Archimedean geometry II,   R. CLUCKERS, J. NICAISE & J. SEBAG (eds)
385   Entropy of hidden Markov processes and connections to dynamical systems,   B. MARCUS, K. PETERSEN & T. WEISSMAN (eds)
386   Independence-friendly logic,   A.L. MANN, G. SANDU & M. SEVENSTER
387   Groups St Andrews 2009 in Bath I,   C.M. CAMPBELL *et al* (eds)
388   Groups St Andrews 2009 in Bath II,   C.M. CAMPBELL *et al* (eds)
389   Random fields on the sphere,   D. MARINUCCI & G. PECCATI
390   Localization in periodic potentials,   D.E. PELINOVSKY
391   Fusion systems in algebra and topology,   M. ASCHBACHER, R. KESSAR & B. OLIVER
392   Surveys in combinatorics 2011,   R. CHAPMAN (ed)
393   Non-abelian fundamental groups and Iwasawa theory,   J. COATES *et al* (eds)
394   Variational problems in differential geometry,   R. BIELAWSKI, K. HOUSTON & M. SPEIGHT (eds)
395   How groups grow,   A. MANN
396   Arithmetic differential operators over the p-adic integers,   C.C. RALPH & S.R. SIMANCA
397   Hyperbolic geometry and applications in quantum chaos and cosmology,   J. BOLTE & F. STEINER (eds)
398   Mathematical models in contact mechanics,   M. SOFONEA & A. MATEI
399   Circuit double cover of graphs,   C.-Q. ZHANG
400   Dense sphere packings: a blueprint for formal proofs,   T. HALES
401   A double Hall algebra approach to affine quantum Schur–Weyl theory,   B. DENG, J. DU & Q. FU
402   Mathematical aspects of fluid mechanics,   J.C. ROBINSON, J.L. RODRIGO & W. SADOWSKI (eds)
403   Foundations of computational mathematics, Budapest 2011,   F. CUCKER, T. KRICK, A. PINKUS & A. SZANTO (eds)
404   Operator methods for boundary value problems,   S. HASSI, H.S.V. DE SNOO & F.H. SZAFRANIEC (eds)
405   Torsors, étale homotopy and applications to rational points,   A.N. SKOROBOGATOV (eds)
406   Appalachian set theory,   J. CUMMINGS & E. SCHIMMERLING (eds)
407   The maximal subgroups of the low-dimensional finite classical groups,   J.N. BRAY, D.F. HOLT & C.M. RONEY-DOUGAL
408   Complexity science: the Warwick master's course,   R. BALL, V. KOLOKOLTSOV & R.S. MACKAY (eds)
409   Surveys in combinatorics 2013,   S.R. BLACKBURN, S. GERKE & M. WILDON (eds)
410   Representation theory and harmonic analysis of wreath products of finite groups,   T. CECCHERINI-SILBERSTEIN, F. SCARABOTTI & F. TOLLI
411   Moduli spaces,   L. BRAMBILA-PAZ, O. GARCÍA-PRADA, P. NEWSTEAD & R.P. THOMAS (eds)
412   Automorphisms and equivalence relations in topological dynamics,   D.B. ELLIS & R. ELLIS
413   Optimal transportation,   Y. OLLIVIER, H. PAJOT & C. VILLANI (eds)
414   Automorphic forms and Galois representations I,   F. DIAMOND, P.L. KASSAEI & M. KIM (eds)
415   Automorphic forms and Galois representations II,   F. DIAMOND, P.L. KASSAEI & M. KIM (eds)
416   Reversibility in dynamics and group theory,   A.G. O'FARRELL & I. SHORT
417   Recent advances in algebraic geometry,   C.D. HACON, M. MUSTAŢĂ & M. POPA (eds)
418   The Bloch–Kato conjecture for the Riemann zeta function,   J. COATES, A. RAGHURAM, A. SAIKIA & R. SUJATHA (eds)
419   The Cauchy problem for non-Lipschitz semi-linear parabolic partial differential equations,   J.C. MEYER & D.J. NEEDHAM
420   Arithmetic and geometry,   L. DIEULEFAIT *et al* (eds)
421   O-minimality and Diophantine geometry,   G.O. JONES & A.J. WILKIE (eds)
422   Groups St Andrews 2013,   C.M. CAMPBELL *et al* (eds)
423   Inequalities for graph eigenvalues,   Z. STANIĆ
424   Surveys in combinatorics 2015,   A. CZUMAJ *et al* (eds)

London Mathematical Society Lecture Note Series: 419

The Cauchy Problem for Non-Lipschitz Semi-Linear Parabolic Partial Differential Equations

J. C. MEYER
University of Birmingham

D. J. NEEDHAM
University of Birmingham

CAMBRIDGE
UNIVERSITY PRESS

CAMBRIDGE
UNIVERSITY PRESS

Shaftesbury Road, Cambridge CB2 8EA, United Kingdom

One Liberty Plaza, 20th Floor, New York, NY 10006, USA

477 Williamstown Road, Port Melbourne, VIC 3207, Australia

314–321, 3rd Floor, Plot 3, Splendor Forum, Jasola District Centre, New Delhi – 110025, India

103 Penang Road, #05–06/07, Visioncrest Commercial, Singapore 238467

Cambridge University Press is part of Cambridge University Press & Assessment, a department of the University of Cambridge.

We share the University's mission to contribute to society through the pursuit of education, learning and research at the highest international levels of excellence.

www.cambridge.org
Information on this title: www.cambridge.org/9781107477391

© J. C. Meyer and D. J. Needham 2015

First published 2015

A catalogue record for this publication is available from the British Library

Library of Congress Cataloging-in-Publication data
Meyer, J. C. (John Christopher)
The Cauchy problem for non-Lipschitz semi-linear parabolic partial differential equations / J.C. Meyer, University of Birmingham, D.J. Needham, University of Birmingham.
pages cm. – (London Mathematical Society lecture note series; 419)
Includes bibliographical references and index.
ISBN 978-1-107-47739-1
1. Cauchy problem.   2. Differential equations, Partial.   3. Differential equations, Parabolic.   I. Needham, D.J. (David J.)   II. Title.
QA377.M494   2015
515´.3534–dc23
2014044865

ISBN   978-1-107-47739-1   Paperback

Contents

v

Notations

$\mathrm{BPC}^2(\mathbb{R})$	The set of bounded, continuous functions $v : \mathbb{R} \to \mathbb{R}$ with continuous derivative and piecewise continuous second derivative
$\mathrm{BPC}_+^2(\mathbb{R})$	The set of functions $v \in \mathrm{BPC}^2(\mathbb{R})$ for which $v : \mathbb{R} \to \mathbb{R}$ is non-negative
$\mathrm{BPC}_{+'}^2(\mathbb{R})$	The set of functions $v \in \mathrm{BPC}_+^2(\mathbb{R})$ for which $v : \mathbb{R} \to \mathbb{R}$ is not equivalently zero
B_A^T	The set of bounded continuous functions $u : \bar{D}_T \to \mathbb{R}$
B_B	The set of bounded continuous functions $v : \mathbb{R} \to \mathbb{R}$
$C^1([0, T])$	The set of continuously differentiable functions $f : [0, T] \to \mathbb{R}$
H_α	The set of locally Hölder continuous functions $f : \mathbb{R} \to \mathbb{R}$
L	The set of locally Lipschitz continuous functions $f : \mathbb{R} \to \mathbb{R}$
$L^1([0, T])$	The set of Lebesgue integrable functions $f : [0, T] \to \mathbb{R}$
L_u	The set of locally upper Lipschitz continuous functions $f : \mathbb{R} \to \mathbb{R}$
(B-D-C)	Bounded diffusion Cauchy problem
(B-R-D-C)	Bounded reaction-diffusion Cauchy problem
(I-B-D-C)	Inhomogeneous bounded diffusion Cauchy problem
(R-S-B)	Regular sub-solution
(R-S-P)	Regular super-solution
(S-R-D-C-1)	(B-R-D-C) with $f(u) = -[u^p]_+$ and $u_0 \in \mathrm{BPC}_+^2(\mathbb{R})$
(S-R-D-C-2)	(B-R-D-C) with $f(u) = [u^p]_+$ and $u_0 \in \mathrm{BPC}_{+'}^2(\mathbb{R})$
(S-R-D-C-3)	(B-R-D-C) with $f(u) = [u^p]_+ + [(1 - u)^q]_+$ and $u_0 \in \mathrm{BPC}_{+'}^2(\mathbb{R})$
$\mathcal{S}$	The set of solutions to (B-R-D-C)

1

Introduction

The study of solutions to systems of semi-linear parabolic partial differential equations has attracted considerable attention over the past fifty years. In the case when the nonlinearity satisfies a local Lipschitz condition, the fundamental theory is well developed (see, for example, the texts of Friedman [21], Fife [20], Rothe [65], Smoller [70], Samarskii *et al.* [67], Volpert *et al.* [72], Leach and Needham [36], and references therein). The situation when the nonlinearity does not necessarily satisfy a local Lipschitz condition is less well studied, but contributions have been made in the case of specific non-Lipschitz nonlinearities which have aided in particular applications (see, for example, Aguirre and Escobedo [5]; Needham *et al.* [36], [54], [29], [33], [40], [41], [42], [43] and references therein), and for the corresponding steady state elliptic problems (see, for example, Stakgold [71], Bandle *et al* [9], [10], [11], [12], [13], Abdullaev [2], [3], [4], [1] and references therein). The aim of this monograph is to exhibit general results concerning semi-linear parabolic partial differential equations that do not necessarily satisfy a local Lipschitz condition. The approach is classical, in the sense that the results relate entirely to the well-posedness criteria for classical solutions, in the sense of Hadamard [39], and the main results are principally established within the framework of real analysis. The approach used to develop the existence theory in this monograph has similarities with the method of successive approximations for systems of first order ordinary differential equations, as detailed in [17] and [16]. Alternative approaches may be possible through the concepts of weak solutions and the framework of semigroup theory. These alternative approaches are amenable, and very effective, in the case of Lipschitz continuous nonlinearities, as exemplified in the monographs by Henry [26] and Pazy [62]. However, the extensions to non-Lipschitz nonlinearities have not been developed and our approach provides an effective development of the classical theory for Lipschitz continuous nonlinearities.

The theory developed in this monograph is applicable to models that arise naturally in many areas of scientific interest. For example, physical, biophysical and environmental modelling gives rise to semi-linear parabolic partial differential equations (often referred to as reaction-diffusion equations) in such areas as population dynamics (see, for example, Levin [37]), the spread of infectious disease (see, for example, Kermack and McKendrick [30], [31] and [32]), smouldering combustion (see, for example, Aris [7] and [8]), isothermal autocatalytic reaction dynamics (see, for example, Gray and Scott [23]), biochemical morphogenesis (see, for example, Murray [52]) and diffusion in complex polymeric materials (see, for example, Edwards [19]). For the purpose of this monograph, we give an introduction based on modelling arising from a chemical kinetics context. Specifically this is motivated by the study of the dynamics of several particular models of autocatalytic chemical reactions under molecular diffusion. These studies can be found in Needham *et al.* [33], [47], [46], [44], [22], [45], [57] and [54]. The aim of this monograph is to develop a generic theory that both encompasses and considerably extends the more specific approaches developed independently in [5], [33] and [54].

The mathematical model concerns the dynamics of an isothermal, autocatalytic chemical reaction scheme with termination, taking place in an unstirred environment and undergoing molecular diffusion. Formally the autocatalytic reaction model is represented by two steps

$$A \longrightarrow B \text{ at rate } k_1 a^q b^p \qquad \text{(autocatalysis)}$$
$$B \longrightarrow C \text{ at rate } k_2 b^r \qquad \text{(decay)}$$

where $p, q, r \in (0, \infty)$ and $k_1, k_2 \in (0, \infty)$ represent the order of each reaction term and the reaction rate constants respectively, whilst a and b represent the concentrations of the reactant A and the autocatalyst B respectively. The chemical C is a stable product of the reaction. At time $\bar{t} = 0$, the autocatalyst is introduced into an expanse of the reactant, which is at uniform concentration $a_0 \geq 0$. This leads to the coupled reaction-diffusion initial boundary value problem, namely,

$$\left.\begin{array}{l} \dfrac{\partial a}{\partial \bar{t}} = D_a \dfrac{\partial^2 a}{\partial \bar{x}^2} - k_1 [a^q]^+ [b^p]^+ \\[2mm] \dfrac{\partial b}{\partial \bar{t}} = D_b \dfrac{\partial^2 b}{\partial \bar{x}^2} + k_1 [a^q]^+ [b^p]^+ - k_2 [b^r]^+ \end{array}\right\} \text{ for } -\infty < \bar{x} < \infty, \ \bar{t} > 0,$$

(1.1)

$$a(\bar{x}, 0) = a_0, \quad b(\bar{x}, 0) = b_0 u_0(\bar{x}) \text{ for } -\infty < \bar{x} < \infty, \qquad (1.2)$$

$a(\bar{x}, \bar{t})$ and $b(\bar{x}, \bar{t})$ are bounded as $|\bar{x}| \to \infty$

uniformly for $0 \leq \bar{t} \leq T$ and any $T \geq 0$. $\qquad$ (1.3)

Here $u_0 : \mathbb{R} \to \mathbb{R}$ is bounded and continuous, with bounded derivative and bounded piecewise continuous second derivative, and

$$\sup_{\bar{x} \in \mathbb{R}} u_0(\bar{x}) = 1, \quad \inf_{\bar{x} \in \mathbb{R}} u_0(\bar{x}) = 0. \qquad (1.4)$$

The function $u_0 : \mathbb{R} \to \mathbb{R}$ represents the initial concentration distribution of the autocatalyst, with the constant $b_0 \geq 0$ measuring the maximum initial concentration of the autocatalyst, whilst $\bar{x}$ and $\bar{t}$ represent the spatial distance and time. The positive constants D_a and D_b represent the diffusion coefficients for species A and B respectively. Here we define the function $[\ ]^+ : \mathbb{R}^2 \to \mathbb{R}$ for any $a, q \in \mathbb{R}$ to be

$$[\ ]^+(a, q) \equiv [a^q]^+ = \begin{cases} a^q & \text{for } a > 0, \\ 0 & \text{for } a \leq 0. \end{cases}$$

When no autocatalysis occurs in (1.1)–(1.3), or equivalently, $k_1 = 0$, then, equation (1.1) and condition (1.2), reduce to

$$a(\bar{x}, \bar{t}) = a_0, \quad \frac{\partial b}{\partial \bar{t}} = D_b \frac{\partial^2 b}{\partial \bar{x}^2} - k_2 [b^r]^+ \quad \text{for } -\infty < \bar{x} < \infty, \ \bar{t} > 0. \quad (1.5)$$

It is convenient to introduce the function $u : \mathbb{R} \times [0, \infty) \to \mathbb{R}$ and dimensionless variables $\tilde{x}$ and $\tilde{t}$, via

$$b(\bar{x}, \bar{t}) = b_0 u(\tilde{x}, \tilde{t}), \quad \tilde{t} = \left(\frac{k_2}{b_0^{(1-r)}}\right) \bar{t}, \quad \tilde{x} = \left(\frac{k_2}{D_b b_0^{(1-r)}}\right)^{1/2} \bar{x}. \quad (1.6)$$

On substituting from (1.6) into (1.5), the system (1.1)–(1.3) becomes

$$\frac{\partial u}{\partial \tilde{t}} = \frac{\partial^2 u}{\partial \tilde{x}^2} - [u^r]^+ \quad \text{for } -\infty < \tilde{x} < \infty, \ \tilde{t} > 0, \qquad (1.7)$$

$$u(\tilde{x}, 0) = u_0(\tilde{x}) \quad \text{for } -\infty < \tilde{x} < \infty, \qquad (1.8)$$

$u(\tilde{x}, \tilde{t})$ is uniformly bounded as $|\tilde{x}| \to \infty$ for each $t \in [0, T]$ and any $T > 0$. (1.9)

The study of the initial-boundary value problem given by (1.7)–(1.9) gives information about the dynamics of the original chemical system in the absence of autocatalysis, which directly motivates the theory of Chapters 6–8 and Chapter 9, Section 9.1. This particular problem has been studied extensively when $r \in [1, \infty)$. However, the case $r \in (0, 1)$ has received much less attention.

Returning to the full system (1.1)–(1.3), when the molecular sizes of A and B are comparable, then we can make the simplification

$$D_a = D_b = D. \tag{1.10}$$

We now introduce the functions $\alpha, \beta : \mathbb{R} \times [0, \infty) \to \mathbb{R}$ and dimensionless variables

$$a(\bar{x}, \bar{t}) = a_0 \alpha(x, t), \quad b(\bar{x}, \bar{t}) = a_0 \beta(x, t),$$

$$\bar{x} = \left(\frac{D}{a_0^{p+q-1} k_1} \right)^{1/2} x, \quad \bar{t} = \frac{1}{a_0^{p+q-1} k_1} t. \tag{1.11}$$

On using (1.10) and (1.11), the initial-boundary value problem (1.1)–(1.3) becomes, in dimensionless form

$$\left. \begin{aligned} \frac{\partial \alpha}{\partial t} &= \frac{\partial^2 \alpha}{\partial x^2} - [\alpha^q]^+ [\beta^p]^+ \\ \frac{\partial \beta}{\partial t} &= \frac{\partial^2 \beta}{\partial x^2} + [\alpha^q]^+ [\beta^p]^+ - k[\beta^r]^+ \end{aligned} \right\} \quad \text{for } -\infty < x < \infty, \ t > 0, \tag{1.12}$$

$$\alpha(x, 0) = 1, \ \beta(x, 0) = \beta_0 u_0(x) \quad \text{for } -\infty < x < \infty, \tag{1.13}$$

$$\alpha(x, t) \text{ and } \beta(x, t) \text{ are bounded as } |x| \to \infty \text{ uniformly}$$
$$\text{for } t \in [0, T] \text{ and any } T > 0. \tag{1.14}$$

Here we have introduced the dimensionless parameters $\beta_0 = \frac{b_0}{a_0}$ and $k = \frac{k_2}{k_1} a_0^{r-p-q}$. Next, we consider the situation in the absence of the termination step, corresponding to $k = 0$ in (1.12). Additionally, when β_0 is small, we can make the approximation

$$\alpha(x, t) + \beta(x, t) = 1 \quad \text{for } -\infty < x < \infty, \ t \geq 0. \tag{1.15}$$

On substituting from (1.15) into the system (1.12)–(1.14), with $k = 0$, and setting $v = \beta$, leads to the reduced scalar problem

$$\frac{\partial v}{\partial t} = \frac{\partial^2 v}{\partial x^2} + [(1 - v)^q]^+ [v^p]^+ \quad \text{for } -\infty < x < \infty, \ t > 0, \tag{1.16}$$

$$v(x, 0) = v_0 u_0(x) \quad \text{for } -\infty < x < \infty, \tag{1.17}$$

$$v(x, t) \text{ is uniformly bounded as } |x| \to \infty \text{ for } t \in [0, T] \text{ and any } T > 0, \tag{1.18}$$

where $v_0 = \beta_0$. The study of problem (1.16)–(1.18) gives information about the dynamics of the original chemical system in the absence of termination and motivates the theory of Chapter 6, Chapter 8 and Chapter 9, Section 9.3. Furthermore, when we replace the term $(1 - v)^q$ by 1, we obtain the problem which is an approximation to (1.16)–(1.18) when v_0 is small, and this motivates the theory in Chapter 9, Section 9.2.

In this monograph, Chapters 2–5 contain essential introductory material, with new specific extensions which are crucial to later chapters. Chapter 2 is a detailed problem statement concerning the class of problems which this monograph examines, as well as a means of introducing the notation and concepts used throughout. Results are also included to highlight the relationship between certain aspects of interest. Chapter 3 contains maximum principles and contains an extension to the classical parabolic maximum principle, as well as some counterexamples to potentially more general results. Moreover, a result is included to illustrate a violation of the strong maximum principle found in [64] and a suitable amendment is included. Chapter 4 is a summary of well established results regarding the theory for solving *diffusion problems* on $\mathbb{R}$ with bounded initial data. Estimates are also provided which are used in the asymptotic results in Chapter 9, Section 9.2. Chapter 5 introduces the integral equations that arise in the study of *reaction-diffusion problems* in later chapters, together with a class of new "Schauder" type derivative estimates.

Chapters 6–8 provide the general results of the monograph. Chapter 6 is largely a review of the question of well-posedness for the reaction-diffusion problem where the reaction function is *Lipschitz continuous*. Chapters 7 and 8 concern the reaction-diffusion problem when the reaction function is *upper Lipschitz continuous* and *Hölder continuous* respectively. In both of these chapters, conditional well-posedness results are established.

Chapter 9 is dedicated to the study of the three specific problems. The study of the well-posedness and qualitative behavior of solutions to these problems are dealt with for the most part by the theory developed in Chapters 6–8. However the problems in Chapter 9 cannot be fully dealt with by the general theory and problem specific results have also been established.

Chapter 10 discusses possible extensions to the theory developed in the monograph and poses open questions which have arisen through the studies in the monograph.

Throughout the monograph, previously established results that have been proved in the context of the monograph by the authors are marked with (†). When this has not been deemed necessary, instruction or reference to a proof is supplied. All new results are marked with (‡).

2

The Bounded Reaction-Diffusion Cauchy Problem

We begin by introducing the regions in which the forthcoming initial value problems will be defined. Here $T > 0$, $\delta \in [0, T)$ and $X > 0$ and the following sets are introduced:

$$D_T = (-\infty, \infty) \times (0, T],$$

$$\bar{D}_T = (-\infty, \infty) \times [0, T],$$

$$\partial D = (-\infty, \infty) \times \{0\},$$

$$\bar{D}_T^\delta = (-\infty, \infty) \times [\delta, T],$$

$$D_T^{\delta, X} = (-X, X) \times (\delta, T],$$

$$\bar{D}_T^{\delta, X} = [-X, X] \times [\delta, T],$$

$$\partial D^{\delta, X} = [-X, X] \times \{\delta\}.$$

The content of the monograph concerns the study of classical solutions $u : \bar{D}_T \to \mathbb{R}$ to the following semi-linear parabolic Cauchy problem;

$$u_t = u_{xx} + f(u); \quad \forall (x, t) \in D_T, \tag{2.1}$$

$$u(x, 0) = u_0(x); \quad \forall x \in \mathbb{R}, \tag{2.2}$$

$$u(x, t) \text{ is uniformly bounded as } |x| \to \infty \text{ for } t \in [0, T]. \tag{2.3}$$

Here, the reaction function $f : \mathbb{R} \to \mathbb{R}$ is prescribed, and the initial data $u_0 : \mathbb{R} \to \mathbb{R}$ is contained in one of the following classes of functions. Firstly, the set of functions $u_0 : \mathbb{R} \to \mathbb{R}$ which are bounded, continuous, with bounded and continuous derivative, and bounded and piecewise continuous second derivative, which is denoted as

$$\text{BPC}^2(\mathbb{R}).$$

7

Secondly and thirdly, the two subsets of $BPC^2(\mathbb{R})$, the first of which contains only non-negative functions $u_0 \in BPC^2(\mathbb{R})$, and the second, which contains only non-negative functions $u_0 \in BPC^2(\mathbb{R})$ but excludes the zero function, which are denoted, respectively, by

$$BPC^2_+(\mathbb{R}) \text{ and } BPC^2_{+'}(\mathbb{R}).$$

The partial differential equation (PDE) (2.1) is generally referred to as a *reaction-diffusion equation*, and the initial value problem given by (2.1)–(2.3) will be referred to throughout the monograph as the *bounded, reaction-diffusion Cauchy problem*, abbreviated to (B-R-D-C). Moreover, throughout the monograph, we adopt the following classical definition of solution to (B-R-D-C):

Definition 2.1 A solution to (B-R-D-C) is a function $u : \bar{D}_T \to \mathbb{R}$ which is continuous and bounded on $\bar{D}_T$ and for which u_t, u_x and u_{xx} exist and are continuous on D_T. Moreover $u : \bar{D}_T \to \mathbb{R}$ must satisfy each of (2.1)–(2.3). ⌐

The questions addressed in this monograph concern the global well-posedness of (B-R-D-C) in the sense of Hadamard [39]. In particular, for a given $f : \mathbb{R} \to \mathbb{R}$, we seek to establish,

(P1) (Existence.) for each $u_0 \in \mathcal{A} \subset BPC^2(\mathbb{R})$, there exists a solution $u : \bar{D}_T \to \mathbb{R}$ to (B-R-D-C) on $\bar{D}_T$ for each $T > 0$,

(P2) (Uniqueness.) whenever $u : \bar{D}_T \to \mathbb{R}$ and $v : \bar{D}_T \to \mathbb{R}$ are solutions to (B-R-D-C) on $\bar{D}_T$ for the same $u_0 \in \mathcal{A} \subset BPC^2(\mathbb{R})$, then $u = v$ on $\bar{D}_T$ for each $T > 0$,

(P3) (Continuous Dependence.) given that (P1) and (P2) are satisfied for (B-R-D-C), then given any $u_0' \in \mathcal{A} \subset BPC^2(\mathbb{R})$ and $\epsilon > 0$, there exists a $\delta > 0$ (which may depend on u_0', T and ϵ) such that for all $u_0 \in \mathcal{A} \subset BPC^2(\mathbb{R})$, then

$$\sup_{x \in \mathbb{R}} |u_0(x) - u_0'(x)| < \delta \implies \sup_{(x,t) \in \bar{D}_T} |u'(x,t) - u(x,t)| < \epsilon$$

where $u : \bar{D}_T \to \mathbb{R}$ and $u' : \bar{D}_T \to \mathbb{R}$ are the solutions to (B-R-D-C) corresponding respectively to $u_0, u_0' \in \mathcal{A} \subset BPC^2(\mathbb{R})$. This must hold for each $T > 0$.

When the above three properties (P1)–(P3) are satisfied by (B-R-D-C), then (B-R-D-C) is said to be *globally well-posed* on $\mathcal{A}$. Moreover, when (P1)–(P3) are satisfied by (B-R-D-C) and the constant δ in (P3) depends only on u_0' and ϵ (that is, being independent of T), then (B-R-D-C) is said to be *uniformly globally well-posed* on $\mathcal{A}$. When one or more of the properties (P1)–(P3) are not satisfied, then (B-R-D-C) is said to be *ill-posed* on $\mathcal{A}$. In addition to well-posedness, we shall address some fundamental qualitative features of solutions to (B-R-D-C).

In conjunction with solutions, we introduce two concepts which will be used throughout the monograph.

Definition 2.2 Let $\bar{u}, \underline{u} : \bar{D}_T \to \mathbb{R}$ be continuous on $\bar{D}_T$ and such that $\underline{u}_t, \underline{u}_x, \underline{u}_{xx}, \bar{u}_t, \bar{u}_x$, and $\bar{u}_{xx}$ exist and are continuous on D_T. Suppose further that

$$N[\bar{u}] \equiv \bar{u}_t - \bar{u}_{xx} - f(\bar{u}) \geq 0 \text{ on } D_T,$$

$$N[\underline{u}] \equiv \underline{u}_t - \underline{u}_{xx} - f(\underline{u}) \leq 0 \text{ on } D_T,$$

$$\underline{u}(x,0) \leq u_0(x) \leq \bar{u}(x,0); \quad \forall x \in \mathbb{R},$$

$\underline{u}$ and $\bar{u}$ are uniformly bounded as $|x| \to \infty$ for t $\in [0, T]$.

Then on $\bar{D}_T$, $\underline{u}$ is called a *regular sub-solution* (R-S-B) and $\bar{u}$ is called a *regular super-solution* (R-S-P) to (B-R-D-C). ⌐

In addition, we require the concept of (B-R-D-C) being *a priori bounded*. This is formalised in the following definition.

Definition 2.3 Suppose that, for (B-R-D-C), we can exhibit a constant $l_T > 0$ for each $0 \leq T \leq T^*$ (and some $T^* > 0$) which depends only upon T and $\sup_{x \in \mathbb{R}} |u_0(x)|$, and which is non-decreasing in $0 \leq T \leq T^*$. Suppose, furthermore, that if $u : \bar{D}_T \to \mathbb{R}$ is any solution to (B-R-D-C) on $\bar{D}_T$, then it can be demonstrated that

$$\sup_{(x,t) \in \bar{D}_T} |u(x,t)| \leq l_T,$$

for each $0 \leq T \leq T^*$. We say that (B-R-D-C) is *a priori bounded* on $\bar{D}_T$ for each $0 \leq T \leq T^*$, with bound l_T. ⌐

In (B-R-D-C), the function $f : \mathbb{R} \to \mathbb{R}$ is referred to as the reaction function, and throughout the monograph we will restrict attention to those reaction functions f from one or more of the following classes of functions. The first class of functions is defined as,

Definition 2.4 A function $f : \mathbb{R} \to \mathbb{R}$ is said to be *Lipschitz continuous* if for any closed bounded interval $E \subset \mathbb{R}$ there exists a constant $k_E > 0$ such that for all $x, y \in E$,

$$|f(x) - f(y)| \le k_E |x - y|.$$

The set of all functions $f : \mathbb{R} \to \mathbb{R}$ which satisfy this definition will be denoted by L. ⌐

For example, any differentiable function $f : \mathbb{R} \to \mathbb{R}$ which has bounded derivative on every closed bounded interval $E \subset \mathbb{R}$ is such that $f \in L$. It is also clear that every function $f \in L$ is continuous. This class of functions has been mentioned first, due to the classical theory of bounded reaction-diffusion Cauchy problems being largely restricted to the case of $f \in L$. The second class of functions which we introduce is parameterised by a real number $\alpha \in (0, 1]$ and is defined as,

Definition 2.5 A function $f : \mathbb{R} \to \mathbb{R}$ is said to be *Hölder continuous of degree* $\alpha \in (0, 1]$ if for any closed bounded interval $E \subset \mathbb{R}$ there exists a constant $k_E > 0$ such that for all $x, y \in E$,

$$|f(x) - f(y)| \le k_E |x - y|^{\alpha}.$$

The set of all functions $f : \mathbb{R} \to \mathbb{R}$ which satisfy this definition will be denoted by H_α (note that $H_1 = L$). ⌐

Any differentiable function $f : \mathbb{R} \to \mathbb{R}$ which has bounded derivative on every closed bounded interval $E \subset \mathbb{R}$ is contained in H_α (for every $\alpha \in (0, 1)$) and every function in H_α is continuous (for any $\alpha \in (0, 1)$). The third class of functions which will be considered in the monograph is defined as,

Definition 2.6 A function $f : \mathbb{R} \to \mathbb{R}$ is said to be *upper Lipschitz continuous* if f is continuous, and for any closed bounded interval $E \subset \mathbb{R}$, there exists a constant $k_E > 0$ such that for all $x, y \in E$, with $y \ge x$,

$$f(y) - f(x) \le k_E (y - x).$$

The set of all functions $f : \mathbb{R} \to \mathbb{R}$ which satisfy this definition will be denoted by L_u. ⌐

Also, for any closed bounded interval $E \subset \mathbb{R}$, any $f \in L_u$ is bounded, and, in particular, setting $E = [a, b]$, we have

$$f(b) + k_E(x - b) \le f(x) \le f(a) + k_E(x - a); \quad \forall x \in [a, b]. \tag{2.4}$$

The following will be useful elementary properties associated with functions $f \in L_u$.

Proposition 2.7 *When $f : \mathbb{R} \to \mathbb{R}$ is a continuous and non-increasing function, then $f \in L_u$. Moreover, on any closed bounded interval $E \subset \mathbb{R}$, Definition 2.6 is satisfied by $f : \mathbb{R} \to \mathbb{R}$ for any $k_E > 0$.*

Proof Let $f : \mathbb{R} \to \mathbb{R}$ be a continuous and non-increasing function. Then on any closed bounded interval $E \subset \mathbb{R}$, for any $x, y \in E$ such that $y \geq x$,

$$f(y) - f(x) \leq 0 \leq k_E(y - x), \tag{2.5}$$

for any $k_E > 0$. It follows via Definition 2.6 that $f \in L_u$, as required. □

We also have,

Proposition 2.8 *Let $f \in L_u$, then on every closed bounded interval $E \subset \mathbb{R}$ there exists a constant $k_E > 0$ such that for all $x, y \in E$ with $x \neq y$,*

$$\frac{(f(y) - f(x))}{(y - x)} \leq k_E.$$

Proof Let $f \in L_u$, then there exists a constant $k_E > 0$ such that

$$(f(v) - f(u)) \leq k_E(v - u) \tag{2.6}$$

for all $u, v \in E$ with $v > u$. Now let $x, y \in E$ with $x \neq y$. There are two possibilities.

(i) $\underline{y > x}$. It follows from (2.6) with $v = y$ and $u = x$ that

$$\frac{(f(y) - f(x))}{(y - x)} \leq k_E. \tag{2.7}$$

(ii) $\underline{y < x}$. It follows from (2.6) with $v = x$ and $u = y$ that

$$\frac{(f(x) - f(y))}{(x - y)} \leq k_E$$

which gives

$$\frac{(f(y) - f(x))}{(y - x)} \leq k_E. \tag{2.8}$$

Then it follows from (2.7) and (2.8) that

$$\frac{(f(y) - f(x))}{(y - x)} \leq k_E$$

for all $x, y \in E$ with $x \neq y$, as required. □

We now consider elementary containment relations between the sets L, H_α and L_u. We have

Proposition 2.9 *The following containment relations hold, for any* $\alpha \in (0, 1)$:

(a) $L \subset H_\alpha$;
(b) $L \subset L_u$;
(c) $L_u \nsubseteq H_\alpha$;
(d) $H_\alpha \nsubseteq L_u$.

Proof For any fixed $\alpha \in (0, 1)$, let $\hat{f}_\alpha : \mathbb{R} \to \mathbb{R}$ be given by

$$\hat{f}_\alpha(x) = [x^\alpha]^+$$

and so we may write

$$\hat{f}_\alpha(x) = \alpha \int_0^x g(s)ds; \quad \forall x \in \mathbb{R} \qquad (2.9)$$

with

$$g(x) = \begin{cases} x^{(\alpha-1)} & ; x > 0 \\ 0 & ; x \leq 0 \end{cases}$$

and the improper integral is implied in (2.9). It follows from (2.9) that for any $x, y \in \mathbb{R}$,

$$\left| \hat{f}_\alpha(y) - \hat{f}_\alpha(x) \right| = \alpha \int_{\min\{x,y\}}^{\max\{x,y\}} g(s)ds \leq \alpha \int_0^{|y-x|} s^{(\alpha-1)}ds = |y-x|^\alpha.$$

Thus $\hat{f}_\alpha \in H_\alpha$. However, take $y > 0$, then

$$\frac{\hat{f}_\alpha(y) - \hat{f}_\alpha(0)}{y} = y^{(\alpha-1)},$$

which is unbounded as $y \to 0^+$. Hence $\hat{f}_\alpha \notin L$ and $\hat{f}_\alpha \notin L_u$ via Proposition 2.8. We can now establish (a)–(d).

(a) Let $f \in L$ and $E \subset \mathbb{R}$ be a closed bounded interval $E = [a, b]$. Then, via Definition 2.4, for all $x, y \in E$,

$$|f(y)-f(x)| \leq k_E|y-x| = k_E|y-x|^{(1-\alpha)}|y-x|^\alpha \leq k_E(b-a)^{(1-\alpha)}|y-x|^\alpha.$$

Hence $f \in H_\alpha$. However, $\hat{f}_\alpha \in H_\alpha$ and $\hat{f}_\alpha \notin L$. Thus $L \subset H_\alpha$.

(b) Let $f \in L$ and $E \subset \mathbb{R}$ be a closed bounded interval $E = [a, b]$. Then f is continuous and for all $x, y \in E$ with $y > x$,

$$f(y) - f(x) \leq |f(y) - f(x)| \leq k_E |y - x| = k_E (y - x)$$

and so $f \in L_u$. Now let $H : \mathbb{R} \to \mathbb{R}$ be such that $H(x) = -\hat{f}_{1/2}(x)$ for all $x \in \mathbb{R}$. Then $H \notin L$ but $H \in L_u$, via Proposition 2.7, and so $L \subset L_u$.

(c) Let $G : \mathbb{R} \to \mathbb{R}$ be such that

$$G(x) = \begin{cases} 0 & ; x \leq 0 \\ -\left(\dfrac{-1}{\log(x)}\right)^{1/2} & ; 0 < x < 1/2 \\ -\left(\dfrac{-1}{\log(1/2)}\right)^{1/2} & ; x \geq 1/2. \end{cases} \tag{2.10}$$

Then $G \notin H_\alpha$ (for any fixed $\alpha \in (0, 1)$) but $G \in L_u$, via Proposition 2.7. To see this, suppose that $G \in H_\alpha$ for some $\alpha \in (0, 1)$. Then there exists a constant $k > 0$ such that

$$|G(x)| = \left(\frac{-1}{\log(x)}\right)^{1/2} \leq k|x|^\alpha \quad \forall x \in (0, 1/2],$$

which gives

$$k|x|^\alpha (-\log(x))^{1/2} \geq 1 \quad \forall x \in (0, 1/2]. \tag{2.11}$$

However, the left-hand side of (2.11) approaches zero as $x \to 0^+$, and we obtain a contradiction. Hence $G \notin H_\alpha$ for any $\alpha \in (0, 1)$. Thus $L_u \not\subseteq H_\alpha$.

(d) We observe that $\hat{f}_\alpha \in H_\alpha$ but $\hat{f}_\alpha \notin L_u$ and so $H_\alpha \not\subseteq L_u$. $\qquad\square$

We also have the following inclusion,

Proposition 2.10 $H_{\alpha_1} \subset H_{\alpha_2}$ *for all* $0 < \alpha_2 < \alpha_1 < 1$.

Proof Let $f \in H_{\alpha_1}$ and $E \subset \mathbb{R}$ be a closed bounded interval $E = [a, b]$. Then for all $x, y \in E$,

$$|f(y) - f(x)| \leq k_E |y - x|^{\alpha_1} = k_E |y - x|^{\alpha_2} |y - x|^{(\alpha_1 - \alpha_2)}$$
$$\leq k_E |b - a|^{(\alpha_1 - \alpha_2)} |y - x|^{\alpha_2}, \tag{2.12}$$

since $\alpha_1 > \alpha_2$. Thus it follows that $f \in H_{\alpha_2}$. In addition, $\hat{f}_{\alpha_2} \in H_{\alpha_2}$ but $\hat{f}_{\alpha_2} \notin H_{\alpha_1}$. We conclude that $H_{\alpha_1} \subset H_{\alpha_2}$. $\qquad\square$

Finally, we define the following two additional classes of functions.

Definition 2.11 Let $f : \mathbb{R}^2 \to \mathbb{R}$ satisfy the following condition: for any pair of closed bounded intervals $U, A \subset \mathbb{R}$, there exist constants $k_U > 0$ and $k_A > 0$ such that for all $(u_1, \alpha_1), (u_2, \alpha_2) \in U \times A$,

$$|f(u_1, \alpha_1) - f(u_2, \alpha_2)| \le k_U |u_1 - u_2| + k_A |\alpha_1 - \alpha_2|.$$

The set of all functions $f : \mathbb{R}^2 \to \mathbb{R}$ which satisfy the preceding condition is denoted by L'. ⌐

In fact, Definition 2.11 is equivalent to

$$L' = \left\{ f : \mathbb{R}^2 \to \mathbb{R} : f(u, \cdot), f(\cdot, \alpha) \in L \text{ uniformly on} \right.$$
$$\left. \text{compact intervals in } u \text{ and } \alpha \text{ respectively} \right\}.$$

Definition 2.12 Let $f : \mathbb{R}^2 \to \mathbb{R}$ be continuous and satisfy the following conditions: for any pair of closed bounded intervals $U, A \subset \mathbb{R}$, then there exists constants $k_U > 0$ and $k_A > 0$ such that

(a) for all $u_1, u_2 \in U$ with $u_1 > u_2$, then

$$f(u_1, \alpha) - f(u_2, \alpha) \le k_U(u_1 - u_2); \quad \forall \alpha \in A,$$

(b) for all $\alpha_1, \alpha_2 \in A$, then

$$|f(u, \alpha_1) - f(u, \alpha_2)| \le k_A |\alpha_1 - \alpha_2|; \quad \forall u \in U.$$

The set of all functions $f : \mathbb{R}^2 \to \mathbb{R}$ which satisfy the preceding conditions is denoted by L'_u. ⌐

Definition 2.12 implies that $f : \mathbb{R}^2 \to \mathbb{R}$ is upper Lipschitz continuous in $u \in U$, uniformly for $\alpha \in A$, and Lipschitz continuous in $\alpha \in A$ uniformly for $u \in U$.

3

Maximum Principles

3.1 Classical Maximum Principles

The main content in this chapter consists of two extensions to the following classical maximum principles (see, for example, [64] and [21]). These maximum principles extend the classical result, which concerns compact spatial domains, to spatially non-compact domains. The first extensions of this type were obtained by Krzyżański in [34] and developed extensively in the following decades (see, [64] (p.193–194)). In this section we review these classical maximum principles and highlight a specific difference in their hypotheses with an example.

Theorem† 3.1 (Classical Weak Maximum Principle) *Let $u : \bar{D}_T^{0,X} \to \mathbb{R}$ (for some $X > 0$) be bounded, continuous and such that u_t, u_x and u_{xx} all exist and are continuous on $D_T^{0,X}$. Suppose that*

$$u_t - a(x,t)u_x - u_{xx} - h(x,t)u \leq 0 \quad on \ D_T^{0,X}, \tag{3.1}$$

where $h : \bar{D}_T^{0,X} \to \mathbb{R}$ is bounded above and $a : \bar{D}_T^{0,X} \to \mathbb{R}$ has no regularity restrictions. Then, $u \leq 0$ on $([-X, X] \times \{0\}) \cup (\{-X\} \times [0, T]) \cup (\{X\} \times [0, T])$, implies $u \leq 0$ on $\bar{D}_T^{0,X}$.

Proof Since h is bounded above on $\bar{D}_T^{0,X}$ there exists $H > 0$ such that

$$h(x,t) \leq H \text{ on } \bar{D}_T^{0,X}. \tag{3.2}$$

Now let $w : \bar{D}_T^{0,X} \to \mathbb{R}$ be given by

$$w(x,t) = e^{-2Ht}u(x,t) \quad on \ \bar{D}_T^{0,X}. \tag{3.3}$$

Since u is bounded and continuous on $\bar{D}_T^{0,X}$ and u_t, u_x and u_{xx} all exist and are continuous on $D_T^{0,X}$, then w is bounded and continuous on $\bar{D}_T^{0,X}$ and w_t, w_x and w_{xx} all exist and are continuous on $D_T^{0,X}$. Via (3.1) we also have

15

$$w_t - a(x,t)w_x - w_{xx} - (h(x,t) - 2H)w \le 0 \quad \text{on } D_T^{0,X}, \qquad (3.4)$$

$$w \le 0 \text{ on } (\{-X\} \times [0,T]) \cup ([-X,X] \times \{0\}) \cup (\{X\} \times [0,T]). \qquad (3.5)$$

Suppose now that $w \not\le 0$ on $\bar{D}_T^{0,X}$. Then since w is continuous on $\bar{D}_T^{0,X}$, which is compact, via (3.5) there exists $(x^*, t^*) \in D_T^{0,X}$ such that

$$\sup_{(x,t) \in \bar{D}_T^{0,X}} w = w(x^*, t^*) = M > 0. \qquad (3.6)$$

Moreover, via (3.4), (3.2) and (3.6), we have

$$w_t(x^*, t^*) - a(x^*, t^*)w_x(x^*, t^*) - w_{xx}(x^*, t^*)$$
$$\le (h(x^*, t^*) - 2H)w(x^*, t^*) < 0. \qquad (3.7)$$

There are now two possibilities:

(i) If $t^* \ne T$ then $w_t(x^*, t^*) = w_x(x^*, t^*) = 0$ and $w_{xx}(x^*, t^*) \le 0$. However, via (3.7), we have $w_{xx}(x^*, t^*) > w_t(x^*, t^*) - a(x^*, t^*)w_x(x^*, t^*) = 0$ and we arrive at a contradiction.

(ii) If $t^* = T$ then $w_x(x^*, t^*) = 0$, $w_{xx}(x^*, t^*) \le 0$ and $w_t(x^*, t^*) \ge 0$. However, via (3.7), $w_t(x^*, t^*) < w_{xx}(x^*, t^*) + a(x^*, t^*)w_x(x^*, t^*) \le 0$ and we arrive at a contradiction.

These two cases are the only possibilities and each leads to a contradiction. We conclude that $w \le 0$ on $\bar{D}_T^{0,X}$. Therefore, via (3.3), $u \le 0$ on $\bar{D}_T^{0,X}$, as required. $\qquad \square$

A strong version of Theorem 3.1 originally obtained by Nirenberg [58] has been exhibited in [64] (p.163–172), namely,

Theorem 3.2 (Classical Strong Maximum Principle) *Let* $u : \bar{D}_T^{0,X} \to \mathbb{R}$ *satisfy the regularity conditions of Theorem 3.1 whilst* $a : \bar{D}_T^{0,X} \to \mathbb{R}$ *and* $h : \bar{D}_T^{0,X} \to \mathbb{R}$ *are bounded. Suppose that* $u \le 0$ *on* $([-X, X] \times \{0\}) \cup (\{-X\} \times [0,T]) \cup (\{X\} \times [0,T])$, *then*

(i) $u < 0$ *on* $D_T^{0,X}$,
   or
(ii) $u = 0$ *on* $[-X, X] \times [0, t^*]$, *where*

$$t^* = \sup\{t \in (0,T] : \exists\, x \in (-X, X) \text{ such that } u(x,t) = 0\}.$$

Proof The proof is lengthy and technical and can be found in [64] (p.159–172). $\qquad \square$

Remark 3.3 In [64] Theorem 3.2 is stated with the additional condition that h is non-positive, however this condition has been dropped as the approach used in (3.3) can be applied to obtain a corresponding differential inequality (in this case for w) where $(h - 2H) : \mathbb{R} \to \mathbb{R}$ is non-positive. We also note that in the statement of the Strong Maximum Principle in [64], h is not required to be bounded below (this is not the case in [58]). This is an error, as the following counter example demonstrates. ⌐

Example‡ 3.4 Consider the function $I : [0, \sigma] \to \mathbb{R}$, with

$$\sigma = \frac{2 + \alpha}{1 + \alpha} > 1, \tag{3.8}$$

and $\alpha \in (0, 1)$, given by

$$I(y) = \int_1^y \frac{1}{s^{(1+\alpha)/2}(\sigma - s)^{1/2}} ds; \quad \forall y \in (0, \sigma), \tag{3.9}$$

with

$$I(0) = \lim_{y \to 0^+} I(y) = I_0 \ (< 0), \quad I(\sigma) = \lim_{y \to \sigma^-} I(y) = I_\sigma \ (> 0). \tag{3.10}$$

It is readily established that I is continuous and bounded on $[0, \sigma]$ and differentiable on $(0, \sigma)$ with derivative given by

$$I'(y) = \frac{1}{y^{(1+\alpha)/2}(\sigma - y)^{1/2}}; \quad \forall y \in (0, \sigma). \tag{3.11}$$

Moreover, (3.11) implies that I is strictly increasing for all $y \in [0, \sigma]$ and hence

$$I : [0, \sigma] \to [I_0, I_\sigma] \text{ is a bijection.} \tag{3.12}$$

It follows from (3.11), (3.12) and the *Inverse function theorem* [66] (p.221–222) that there exists a function $J : [I_0, I_\sigma] \to [0, \sigma]$ such that

$$J(I(y)) = y \quad \forall y \in [0, \sigma], \quad I(J(x)) = x \quad \forall x \in [I_0, I_\sigma],$$
$$J(I_0) = 0, \quad J(I_\sigma) = \sigma. \tag{3.13}$$

It follows from (3.13) that

$$J(I(1)) = J(0) = 1, \quad J(I(0)) = J(I_0) = 0, \quad J(I(\sigma)) = J(I_\sigma) = \sigma. \tag{3.14}$$

Moreover, J is continuous on $[I_0, I_\sigma]$ and differentiable on $[I_0, I_\sigma]$ with derivative given by

$$J'(x) = J(x)^{(1+\alpha)/2}(\sigma - J(x))^{1/2}; \quad \forall x \in [I_0, I_\sigma]. \tag{3.15}$$

It follows from (3.15) that

$$J'(I_0) = J'(I_\sigma) = 0, \quad J'(x) > 0; \; \forall x \in (I_0, I_\sigma). \tag{3.16}$$

Therefore, via (3.15) and (3.12), J is increasing and J' is continuous for $x \in [I_0, I_\sigma]$. Now it follows from the chain rule that, J'' exists for $x \in (I_0, I_\sigma)$, and via (3.8), is given by

$$\begin{aligned}
J''(x) &= \left(\frac{1}{2}\right) J'(x) \left((1+\alpha) J(x)^{(\alpha-1)/2} (\sigma - J(x))^{1/2} \right. \\
&\quad \left. - J(x)^{(\alpha+1)/2} (\sigma - J(x))^{-1/2} \right) \\
&= \frac{(2+\alpha)}{2} J^\alpha(x)(1 - J(x)); \quad \forall x \in (I_0, I_\sigma). \tag{3.17}
\end{aligned}$$

Moreover, since J is continuous, it follows from (3.17) and (3.14) that

$$\lim_{x \to I_0^+} J''(x) = 0, \quad \lim_{x \to I_\sigma^-} J''(x) = -\frac{1}{2}\sigma^{(\alpha+1)}. \tag{3.18}$$

It follows from (3.14)–(3.18) that

$$J : [I_0, I_\sigma] \to \mathbb{R} \text{ is twice continuously differentiable on } [I_0, I_\sigma], \tag{3.19}$$

with

$$J''(I_0) = 0, \quad J''(I_\sigma) = -\frac{1}{2}\sigma^{(\alpha+1)}, \tag{3.20}$$

and so, via (3.17) and (3.20),

$$J''(x) = \frac{(2+\alpha)}{2} J^\alpha(x)(1 - J(x)); \quad \forall x \in [I_0, I_\sigma]. \tag{3.21}$$

We now introduce the function $\tilde{J} : \mathbb{R} \to [0, \sigma]$, given as

$$\tilde{J}(x) = \begin{cases} J(x + I_\sigma) & ; x \in [I_0 - I_\sigma, 0] \\ J(I_\sigma - x) & ; x \in [0, I_\sigma - I_0] \\ 0 & ; x \in \mathbb{R} \setminus [I_0 - I_\sigma, I_\sigma - I_0]. \end{cases} \tag{3.22}$$

Observe that via (3.22), (3.14) and (3.16),

$$\tilde{J} \text{ is continuously differentiable on } \mathbb{R}. \tag{3.23}$$

Moreover, since $\tilde{J} : \mathbb{R} \to [0, \sigma]$ is an even function, it follows from (3.19), (3.20), (3.21) and (3.22) that

$$\tilde{J} \text{ is twice continuously differentiable on } \mathbb{R} \tag{3.24}$$

and

$$\tilde{J}''(x) = \frac{(2+\alpha)}{2} \tilde{J}^\alpha(x)(1 - \tilde{J}(x)); \quad \forall x \in \mathbb{R}. \tag{3.25}$$

Now we introduce the function $\hat{J} : \mathbb{R} \to [0, \sigma]$ given by

$$\hat{J}(x) = \tilde{J}\left(\left(\frac{2}{2+\alpha}\right)^{1/2} x\right); \quad \forall x \in \mathbb{R}. \tag{3.26}$$

It follows from (3.24) that $\hat{J}$ is twice continuously differentiable on $\mathbb{R}$ and via (3.25),

$$\hat{J}''(x) = \hat{J}^\alpha(x)(1 - \hat{J}(x)); \quad \forall x \in \mathbb{R}. \tag{3.27}$$

Now observe, via (3.9), (3.10) and (3.26), that $\hat{J} : \mathbb{R} \to [0, \sigma]$ satisfies

$$\hat{J}(x) > 0; \quad \forall x \in (-X(\alpha), X(\alpha)), \quad \hat{J}(x) = 0; \quad \forall x \in \mathbb{R} \setminus (-X(\alpha), X(\alpha)), \tag{3.28}$$

where

$$\begin{aligned}
X(\alpha) &= \left(\frac{(2+\alpha)}{2}\right)^{1/2} (I_\sigma - I_0) \\
&= \left(\frac{(2+\alpha)}{2}\right)^{1/2} \int_0^\sigma \frac{1}{s^{(1+\alpha)/2}(\sigma - s)^{1/2}} ds \\
&= \left(\frac{(2+\alpha)}{2}\right)^{1/2} \left(\frac{1}{\sigma^{\alpha/2}}\right) \int_0^1 \frac{1}{w^{(1+\alpha)/2}(1 - w)^{1/2}} dw \\
&= \frac{(2+\alpha)^{(1-\alpha)/2}(1+\alpha)^{\alpha/2}}{2^{1/2}} \int_0^1 \frac{1}{w^{(1+\alpha)/2}(1 - w)^{1/2}} dw. \tag{3.29}
\end{aligned}$$

Next define $u : \bar{D}_1^{0,2X(\alpha)} \to [-\sigma, 0]$ such that

$$u(x, t) = -\hat{J}(x); \quad \forall (x, t) \in \bar{D}_1^{0,2X(\alpha)}, \tag{3.30}$$

and so

$$u(x, t) \leq 0; \quad \forall (x, t) \in ([-2X(\alpha), 2X(\alpha)] \times \{0\}) \cup (\{-2X(\alpha), 2X(\alpha)\} \times (0, 1]). \tag{3.31}$$

Moreover, we observe, via (3.28), that

$$u(x, t) < 0; \quad \forall (x, t) \in (-X(\alpha), X(\alpha)) \times [0, 1],$$

$$u(x, t) = 0; \quad \forall (x, t) \in \bar{D}_1^{0,2X(\alpha)} \setminus ((-X(\alpha), X(\alpha)) \times [0, 1]). \tag{3.32}$$

Also, via (3.27) and (3.30), u_t, u_x and u_{xx} exist and are continuous on $D_1^{0,2X(\alpha)}$, and u satisfies

$$u_t - u_{xx} - h(x, t)u = 0 \quad \text{on } D_1^{0,2X(\alpha)}. \tag{3.33}$$

Here $h : \bar{D}_1^{0,2X(\alpha)} \to \mathbb{R}$ is given by

$$
h(x,t) = \begin{cases} \hat{J}(x)^{\alpha-1}(\hat{J}(x) - 1) & ; (x,t) \in (-X(\alpha), X(\alpha)) \times [0,1] \\ 0 & ; (x,t) \in \bar{D}_1^{0,2X(\alpha)} \setminus ((-X(\alpha), X(\alpha)) \times [0,1]) . \end{cases}
$$

(3.34)

We now observe that since $\hat{J}(x) \in (0, \sigma]$ for all $x \in (-X(\alpha), X(\alpha))$, then it follows from (3.34) that

$$
h(x,t) \le \sigma - 1; \quad \forall (x,t) \in \bar{D}_1^{0,2X(\alpha)} . \tag{3.35}
$$

However, h is not bounded below on $\bar{D}_1^{0,2X(\alpha)}$. In addition

$$
t^* = \sup \{t \in (0,1] : \exists\, x \in (-2X(\alpha), 2X(\alpha)) \text{ such that } u(x,t) = 0\} = 1,
$$

whilst

$$
u(x,t) < 0; \quad \forall (x,t) \in (-X(\alpha), X(\alpha)) \times [0,1].
$$

Thus $u : \bar{D}_1^{0,2X(\alpha)} \to \mathbb{R}$ satisfies all the conditions of the Strong Maximum Principle (Theorem 3.2) except that h is not bounded below on $\bar{D}_T^{0,2X(\alpha)}$ in (3.33). We have demonstrated that u violates the conclusions of the Strong Maximum Principle (Theorem 3.2) on $\bar{D}_T^{0,2X(\alpha)}$. We conclude that the Strong Maximum Principle does not hold, in general, when h is not bounded below. ⌐

3.2 Extended Maximum Principles

In this section we now introduce the two extensions to the classical maximum principles (as given in Theorem 3.1 and Theorem 3.2). Extensions of this type are also contained in Chapter 2 of [21] and referenced in [64] (p.193–194). However, the specific restrictions on these results are significantly different in what follows.

Theorem† 3.5 (Extended Maximum Principle 1) *Let* $u : \bar{D}_T \to \mathbb{R}$ *be bounded, continuous and such that* u_t, u_x *and* u_{xx} *all exist and are continuous on* D_T *and* $u(x,t) \to l(\le 0)$ *as* $|x| \to \infty$ *uniformly for* $t \in [0,T]$. *Suppose that*

$$
u_t - a(x,t)u_x - u_{xx} - h(x,t)u \le 0 \text{ on } D_T \tag{3.36}
$$

where $h : \bar{D}_T \to \mathbb{R}$ *is bounded above and* $a : \bar{D}_T \to \mathbb{R}$ *has no regularity restrictions. Then* $u \le 0$ *on* ∂D *implies* $u \le 0$ *on* $\bar{D}_T$.

Proof Since h is bounded above on $\bar{D}_T$ there exists $H > 0$ such that

$$h(x, t) \leq H \text{ on } \bar{D}_T. \tag{3.37}$$

Now let $w : \bar{D}_T \to \mathbb{R}$ be given by

$$w(x, t) = e^{-2Ht} u(x, t) \text{ on } \bar{D}_T. \tag{3.38}$$

Note that since u is bounded and continuous on $\bar{D}_T$ and u_t, u_x and u_{xx} all exist and are continuous on D_T, then w is bounded and continuous on $\bar{D}_T$. Furthermore, via the product rule, w_t, w_x and w_{xx} all exist and are continuous on D_T. Via (3.36) we also have

$$w_t - a(x, t)w_x - w_{xx} - (h(x, t) - 2H)w \leq 0 \text{ on } D_T, \tag{3.39}$$

$$w \leq 0 \text{ on } \partial D, \tag{3.40}$$

and

$$w(x, t) \to le^{-2Ht}(\leq 0) \text{ as } |x| \to \infty \text{ uniformly for } t \in [0, T]. \tag{3.41}$$

Suppose now that $w \not\leq 0$ on $\bar{D}_T$. Then since w is bounded and continuous on $\bar{D}_T$ there exist, via (3.41) and (3.40), $(x^*, t^*) \in D_T$ such that

$$\sup_{(x,t) \in \bar{D}_T} w(x, t) = w(x^*, t^*) = M > 0. \tag{3.42}$$

Moreover, via (3.39), (3.37) and (3.42), we have

$$w_t(x^*, t^*) - a(x^*, t^*)w_x(x^*, t^*) - w_{xx}(x^*, t^*)$$
$$\leq (h(x^*, t^*) - 2H)w(x^*, t^*) < 0. \tag{3.43}$$

There are now two possibilities.

(i) If $t^* \neq T$ then $w_t(x^*, t^*) = w_x(x^*, t^*) = 0$ and $w_{xx}(x^*, t^*) \leq 0$. However, via (3.43), we have $w_{xx}(x^*, t^*) > w_t(x^*, t^*) - a(x^*, t^*)w_x(x^*, t^*) = 0$ and we arrive at a contradiction.
(ii) If $t^* = T$ then $w_x(x^*, t^*) = 0$, $w_{xx}(x^*, t^*) \leq 0$ and $w_t(x^*, t^*) \geq 0$. However, via (3.43), $w_t(x^*, t^*) < w_{xx}(x^*, t^*) + a(x^*, t^*)w_x(x^*, t^*) \leq 0$ and we arrive at a contradiction.

These two cases are the only possibilities and each leads to a contradiction. We conclude that $w \leq 0$ on $\bar{D}_T$. Therefore via (3.38), $u \leq 0$ on $\bar{D}_T$, as required. $\qquad \square$

Theorem‡ 3.6 (Extended Maximum Principle 2) *Let* $u : \bar{D}_T \to \mathbb{R}$ *be bounded, continuous and such that* u_t, u_x *and* u_{xx} *all exist and are continuous on* D_T. *Suppose that*

$$u_t - u_{xx} - a(x,t)u_x - h(x,t)u \le 0 \text{ on } D_T \qquad (3.44)$$

where $h : \bar{D}_T \to \mathbb{R}$ *is bounded above and* $a : \bar{D}_T \to \mathbb{R}$ *satisfies*

(i) *For each* $X > 0$ *there exists a constant* $A_X > 0$ *such that* $xa(x,t) \le A_X$ *for all* $(x,t) \in [-X, X] \times [0, T]$.
(ii) *There exist constants* $A_\infty > 0$ *and* $X_\infty > 0$ *such that* $|a(x,t)| \le A_\infty|x|$ *for all* $(x,t) \in ((-\infty, -X_\infty] \cup [X_\infty, \infty)) \times [0, T]$.

Then $u \le 0$ *on* ∂D *implies* $u \le 0$ *on* $\bar{D}_T$.

Proof Let $w : \bar{D}_T \to \mathbb{R}$ be given by

$$w(x,t) = u(x,t)\hat{\phi}(x,t) \text{ on } \bar{D}_T, \qquad (3.45)$$

where $\hat{\phi} : \bar{D}_T \to \mathbb{R}$ is given by

$$\hat{\phi}(x,t) = \frac{1}{1+x^2}, \qquad (3.46)$$

for all $(x,t) \in \bar{D}_T$. We observe that w is bounded and continuous on $\bar{D}_T$, and moreover

$$w(x,t) \to 0 \text{ as } |x| \to \infty, \text{ uniformly for } t \in [0, T]. \qquad (3.47)$$

We also observe that $\hat{\phi}_t$, $\hat{\phi}_x$ and $\hat{\phi}_{xx}$ all exist and are continuous on D_T and it follows that w_t, w_x and w_{xx} all exist and are continuous on D_T. Since $\hat{\phi} > 0$ on $\bar{D}_T$, (3.45) can be re-written as

$$u(x,t) = w(x,t)\phi(x,t) \text{ on } \bar{D}_T, \qquad (3.48)$$

where $\phi(x,t) = 1 + x^2$ on $\bar{D}_T$. We now substitute from (3.48) into (3.44) to obtain the inequality

$$w_t - w_{xx} - \tilde{a}(x,t)w_x - \tilde{h}(x,t)w \le 0 \text{ on } D_T, \qquad (3.49)$$

where $\tilde{a}, \tilde{h} : \bar{D}_T \to \mathbb{R}$ are given by

$$\tilde{a}(x,t) = a(x,t) + \frac{4x}{1+x^2},$$

$$\tilde{h}(x,t) = \frac{2}{1+x^2} + \frac{2xa(x,t)}{1+x^2} + h(x,t) \qquad (3.50)$$

for all $(x, t) \in \bar{D}_T$. We now establish that $\tilde{h}(x, t)$ is bounded above on $\bar{D}_T$. Via condition (ii) we have

$$\tilde{h}(x, t) \leq 2 + 2A_\infty + h(x, t), \tag{3.51}$$

for all $(x, t) \in \bar{D}_T$ with $|x| > X_\infty$. In addition, via (i), (3.50) implies

$$\tilde{h}(x, t) \leq 2 + 2A_{X_\infty} + h(x, t), \tag{3.52}$$

for all $(x, t) \in \bar{D}_T$ with $|x| \leq X_\infty$. Therefore, (3.51) and (3.52) establish that $\tilde{h}$ is bounded above on $\bar{D}_T$. Furthermore, since $u \leq 0$ on ∂D then $w \leq 0$ on ∂D. Using this together with (3.47), (3.49) and (3.50) it follows from Theorem 3.5 that $w \leq 0$ on $\bar{D}_T$ and hence, via (3.48), that $u \leq 0$ on $\bar{D}_T$, as required. $\quad\square$

We now demonstrate that conditions (i) and (ii) in Theorem 3.6 are not merely technical restrictions, via the construction of two counterexamples.

Here, two functions are considered which satisfy an inequality of the type given in (3.44) of Theorem 3.6 but with the restriction on $a(x, t)$ removed. These demonstrate that a version of Theorem 3.6 without restrictions on $a(x, t)$ is not possible.

Example‡ 3.7 Let $u : \bar{D}_1 \to \mathbb{R}$ be defined as

$$u(x, t) = \begin{cases} -1 + \frac{2\sqrt{2}}{(1+t)^{1/2}} e^{-\left(\frac{(x-\ln(t))^2}{4(1+t)}\right)} & ; (x, t) \in D_1 \\ -1 & ; (x, t) \in \partial D. \end{cases} \tag{3.53}$$

It is readily established that u is continuous on $\bar{D}_1$. Moreover, u_t, u_x and u_{xx} all exist and are continuous on D_1, and are given by

$$u_x(x, t) = \frac{-\sqrt{2}(x - \ln(t))}{(1+t)^{3/2}} e^{-\left(\frac{(x-\ln(t))^2}{4(1+t)}\right)}, \tag{3.54}$$

$$u_{xx}(x, t) = \frac{\sqrt{2}}{(1+t)^{3/2}} \left(-1 + \frac{(x - \ln(t))^2}{2(1+t)}\right) e^{-\left(\frac{(x-\ln(t))^2}{4(1+t)}\right)}, \tag{3.55}$$

$$u_t(x, t) = \frac{\sqrt{2}}{(1+t)^{3/2}} \left(-1 + \frac{(x - \ln(t))}{t} + \frac{(x - \ln(t))^2}{2(1+t)}\right) e^{-\left(\frac{(x-\ln(t))^2}{4(1+t)}\right)} \tag{3.56}$$

for all $(x, t) \in D_1$. Furthermore,

$$|u(x, t)| \leq 2\sqrt{2} - 1 \tag{3.57}$$

for all $(x, t) \in \bar{D}_1$ and so u is bounded on $\bar{D}_1$. Additionally,

$$\sup_{x \in \mathbb{R}} u(x, t) = -1 + \frac{2\sqrt{2}}{(1 + t)^{1/2}} \text{ for } t \in (0, 1], \qquad (3.58)$$

$$\inf_{x \in \mathbb{R}} u(x, t) = -1 \text{ for } t \in (0, 1]. \qquad (3.59)$$

We observe that

$$\sup_{x \in \mathbb{R}} u(x, t) \geq 1 \text{ for all } t \in (0, 1], \qquad (3.60)$$

$$\sup_{x \in \mathbb{R}} u(x, 0) = -1. \qquad (3.61)$$

Moreover, via equations (3.54)–(3.56),

$$u_t - u_{xx} + \frac{1}{t} u_x = 0 \qquad (3.62)$$

for all $(x, t) \in D_1$, and so (3.62) corresponds to the inequality (3.44) with

$$a(x, t) = \begin{cases} \frac{-1}{t} & ; (x, t) \in D_1 \\ 0 & ; (x, t) \in \partial D, \end{cases} \qquad (3.63)$$

$$h(x, t) = 0; \quad \forall (x, t) \in \bar{D}_1. \qquad (3.64)$$

Thus we have constructed a function $u : \bar{D}_1 \to \mathbb{R}$, with $a : \bar{D}_1 \to \mathbb{R}$ and $h : \bar{D}_1 \to \mathbb{R}$ as given in (3.63) and (3.64) respectively, so that all the conditions of Theorem 3.6 are satisfied *except* (i) and (ii) on a, and for which Theorem 3.6 fails. ⌟

Remark 3.8 Observe, via (3.53), that in Example 3.7, $u(x, t) \to -1$ as $|x| \to \infty$ for each fixed $t \in [0, 1]$. However,

$$u(x, t) \not\to -1 \text{ as } |x| \to \infty \text{ } uniformly \text{ for } t \in [0, 1].$$

This feature is related to the unboundedness of $a(x, t)$ as $t \to 0^+$ in $\bar{D}_1$ and leads to the resulting failure of Theorem 3.5 in the above example. ⌟

Example‡ 3.9 Let $w : \bar{D}_1 \to \mathbb{R}$ be defined as

$$w(x, t) = \begin{cases} -1 + 2e^{-\left(\frac{1}{\gamma |x|^\gamma} + 1 - t\right)^2} & ; (x, t) \in \bar{D}_1 \setminus (\{0\} \times [0, 1]) \\ -1 & ; \{0\} \times [0, 1], \end{cases} \qquad (3.65)$$

where $\gamma > 0$ is constant. Observe that w is continuous on $\bar{D}_1$ whilst w_t and w_x exist and are continuous on D_1, and are given by

$$w_t(x,t) = \begin{cases} 4e^{-\left(\frac{1}{\gamma|x|^\gamma}+1-t\right)^2}\left(\frac{1}{\gamma|x|^\gamma}+1-t\right) & ; (x,t) \in D_1\backslash(\{0\}\times(0,1]) \\ 0 & ; (x,t) \in \{0\}\times(0,1], \end{cases}$$

$$(3.66)$$

$$w_x(x,t) = \begin{cases} 4e^{-\left(\frac{1}{\gamma|x|^\gamma}+1-t\right)^2}\left(\frac{1}{\gamma|x|^\gamma}+1-t\right)\left(\frac{1}{|x|^\gamma x}\right) & ; (x,t) \in D_1\backslash(\{0\}\times(0,1]) \\ 0 & ; (x,t) \in \{0\}\times(0,1]. \end{cases}$$

$$(3.67)$$

Moreover,

$$|w(x,t)| \le 1; \quad \forall(x,t) \in \bar{D}_1, \tag{3.68}$$

and so w is bounded on $\bar{D}_1$. Also,

$$w(x,t) \to -1 + 2e^{-(1-t)^2} \text{ as } |x| \to \infty \text{ uniformly for } t \in [0,1], \tag{3.69}$$

$$w(x,0) = \begin{cases} -1 + 2e^{-\left(\frac{1}{\gamma|x|^\gamma}+1\right)^2} & ; x \in \mathbb{R}\backslash\{0\} \\ -1 & ; x = 0. \end{cases} \tag{3.70}$$

Now, via (3.66) and (3.67), we observe that

$$w_t - |x|^\gamma x w_x = 0 \quad \text{on } D_1. \tag{3.71}$$

Additionally, we observe that w_{xx} exists and is continuous on D_1 and is given by

$$w_{xx}(x,t) = \begin{cases} 4e^{-\left(\frac{1}{\gamma|x|^\gamma}+1-t\right)^2}\left[2\left(\frac{1}{\gamma|x|^\gamma}+1-t\right)^2\left(\frac{1}{|x|^{\gamma+1}}\right)^2 \right. \\ \left. -\left(\frac{1}{|x|^{\gamma+1}}\right)^2 - \left(\frac{1}{\gamma|x|^\gamma}+1-t\right)\left(\frac{(\gamma+1)}{|x|^{\gamma+2}}\right)\right] & ; (x,t) \in D_1\backslash(\{0\}\times(0,1]) \\ 0 & ; (x,t) \in \{0\}\times(0,1], \end{cases}$$

$$(3.72)$$

and it follows by inspection that w_{xx} is bounded on D_1. Now introduce $u: \bar{D}_1 \to \mathbb{R}$ given as

$$u(x,t) = w(K^{1/2}x,t) - \frac{1}{2}t, \quad \forall(x,t) \in \bar{D}_1, \tag{3.73}$$

where $K > 0$ is given by

$$K = \left(2 \sup_{(x,t)\in D_1} \{|w_{xx}(x,t)|\}\right)^{-1}. \qquad (3.74)$$

Observe that u is continuous on $\bar{D}_1$, whilst u_t, u_x and u_{xx} exist and are continuous on D_1. Now, via (3.70) and (3.73),

$$u(x,0) = w(K^{1/2}x, 0) \le -1 + 2e^{-1} < 0, \quad \forall x \in \mathbb{R}. \qquad (3.75)$$

Similarly, via (3.69) and (3.73),

$$\lim_{|x|\to\infty} u(x,1) = \lim_{|x|\to\infty} w(K^{1/2}x, 1) - \frac{1}{2} = \frac{1}{2}. \qquad (3.76)$$

It follows from (3.76) that u is *not* non-positive for all $(x,t) \in \bar{D}_1$. In addition, via (3.68) and (3.73), we have

$$|u(x,t)| \le \frac{3}{2}: \quad \forall (x,t) \in \bar{D}_1, \qquad (3.77)$$

and so u is bounded on $\bar{D}_1$. Moreover, via (3.73) we have, for $(x,t) \in D_1$,

$$u_t(x,t) - u_{xx}(x,t) - K^{\gamma/2}|x|^\gamma x u_x(x,t) = \left(w_t(X,t) - |X|^\gamma X w_X(X,t)\right)$$
$$- \frac{1}{2} - K w_{XX}(X,t), \qquad (3.78)$$

where $X = K^{1/2}x$. It then follows from (3.71) and (3.74), that

$$u_t - u_{xx} - K^{\gamma/2}|x|^\gamma x u_x \le 0 \quad \text{on } D_1, \qquad (3.79)$$

which corresponds to the inequality (3.44) in Theorem 3.6 with

$$a(x,t) = K^{\gamma/2}|x|^\gamma x; \quad \forall (x,t) \in \bar{D}_1, \qquad (3.80)$$

$$h(x,t) = 0; \quad \forall (x,t) \in \bar{D}_1. \qquad (3.81)$$

Thus, we have constructed a function $u : \bar{D}_1 \to \mathbb{R}$ with $a : \bar{D}_1 \to \mathbb{R}$ and $h : \bar{D}_1 \to \mathbb{R}$ as given in (3.80) and (3.81) respectively, so that all the conditions of Theorem 3.6 are satisfied *except* (ii) on $a(x,t)$, and for which Theorem 3.6 fails. ⌐

Remark 3.10 Observe that in Example 3.9, it is the superlinear growth of $a : \bar{D}_1 \to \mathbb{R}$ as $|x| \to \infty$, given by (3.80), that leads to the resulting failure of Theorem 3.6. It should also be noted that Example 3.9 is an improvement on the example given in [28] (p.17). ⌐

To close this chapter, we also remark that the principal classical weak and strong maximum principles for parabolic partial differential inequalities (corresponding to Theorem 3.1 and Theorem 3.2) are due to Picone [63] and Nirenberg [58] respectively. For related works see [64].

The first extensions to the above classical maximum principles are due to Krzyżański [34] and were extensively investigated (see, for example, [64] (p.193–194)). The approach adopted here is similar to that of Krzyżański's but differs due to his focus on functions $u : \bar{D}_T \to \mathbb{R}$ that satisfy

$$|u(x, t)| \leq A e^{Bx^2}; \quad \forall (x, t) \in \bar{D}_T$$

for some constants $A, B > 0$. The focus on bounded functions in this monograph, leads to different conditions on the coefficients in the differential inequality. For additional discussion regarding extensions to these results, see [49].

4

Diffusion Theory

In this chapter, we suppose that the molecular diffusion process of an inert chemical species U, with concentration u, is taking place in a one-dimensional infinite domain and the diffusion process has been initiated by the introduction of an initial distribution of the concentration of the species U. In dimensionless variables, the evolution of the concentration u over time $t \in [0, T]$ (for given $T > 0$) is determined by the solution of the *bounded Cauchy problem* for the linear diffusion equation:

(i) $u_t = u_{xx}$; $\forall (x, t) \in D_T$,

(ii) $u(x, 0) = u_0(x)$; $\forall x \in \mathbb{R}$,

(iii) $u(x, t)$ is uniformly bounded as $|x| \to \infty$ for $t \in [0, T]$.

Here $u_0 : \mathbb{R} \to \mathbb{R}$ is the prescribed initial concentration distribution of u. Throughout this chapter we consider the situation when $u_0 \in \text{BPC}^2(\mathbb{R})$. We will refer to this Cauchy problem throughout as (B-D-C). A solution to (B-D-C) follows Definition 2.1 with (B-D-C) replacing (B-R-D-C).

The primary purpose of this chapter is to illustrate a classical well-posedness result for (B-D-C). To this end, the following chapter is a review, in the context of the present body of work, of the classical theory of the Cauchy problem for the linear diffusion equation (see, for example, [21] and [56]). It should be noted that all results marked with † are standard and can be found in some form in [21] and [56]. To begin we have,

Theorem† **4.1** (Uniqueness) *The problem (B-D-C) has at most one solution on $\bar{D}_T$ for any $T > 0$.*

Proof Let $u_1, u_2 : \bar{D}_T \to \mathbb{R}$ both be solutions to (B-D-C) with the same initial data $u_0 \in \text{BPC}^2(\mathbb{R})$. First define $w : \bar{D}_T \to \mathbb{R}$ by

$$w(x, t) = u_1(x, t) - u_2(x, t); \quad \forall (x, t) \in \bar{D}_T. \tag{4.1}$$

Since u_1 and u_2 are solutions to (B-D-C) on $\bar{D}_T$, then w is bounded and continuous on $\bar{D}_T$ whilst w_t, w_x and w_{xx} all exist and are continuous on D_T. Moreover,

$$w_t - w_{xx} = (u_{1t} - u_{1xx}) - (u_{2t} - u_{2xx}) = 0 \le 0 \text{ on } D_T \tag{4.2}$$

and

$$w = 0 \le 0 \text{ on } \partial D. \tag{4.3}$$

It then follows immediately from Theorem 3.6 that

$$w = u_1 - u_2 \le 0 \text{ on } \bar{D}_T \tag{4.4}$$

and so

$$u_1 \le u_2 \text{ on } \bar{D}_T. \tag{4.5}$$

It follows via a symmetrical argument that

$$u_2 \le u_1 \text{ on } \bar{D}_T \tag{4.6}$$

and hence, via (4.6) and (4.5), we have

$$u_1 = u_2 \text{ on } \bar{D}_T, \tag{4.7}$$

as required. □

We next consider the function $u : \bar{D}_T \to \mathbb{R}$ given by

$$u(x, t) = \begin{cases} \dfrac{1}{2\sqrt{\pi t}} \displaystyle\int_{-\infty}^{\infty} u_0(s) e^{-\frac{(x-s)^2}{4t}} ds & ; (x, t) \in D_T \\ u_0(x) & ; (x, t) \in \partial D. \end{cases} \tag{4.8}$$

It is readily established that u is well-defined on $\bar{D}_T$, and a simple substitution gives the alternative representation

$$u(x, t) = \frac{1}{\sqrt{\pi}} \int_{-\infty}^{\infty} u_0(x + 2\sqrt{t}w) e^{-w^2} dw; \quad \forall (x, t) \in \bar{D}_T. \tag{4.9}$$

The boundedness and regularity conditions on $u_0 : \mathbb{R} \to \mathbb{R}$ establish (via classical uniform convergence results; see for example [6]) through (4.9) that u is continuous on $\bar{D}_T$. Moreover, through (4.8), it is further established that u_t, u_x and u_{xx} all exist and are continuous on D_T (via integration by parts and classical uniform convergence results; see for example [6]) and can be obtained via differentiation under the integral sign in (4.8), so that, following integration by parts,

$$u_x(x, t) = \frac{1}{2\sqrt{\pi t}} \int_{-\infty}^{\infty} u_0'(s) e^{-\frac{(x-s)^2}{4t}} ds$$

$$= \frac{1}{\sqrt{\pi}} \int_{-\infty}^{\infty} u_0'(x + 2\sqrt{t}w) e^{-w^2} dw; \quad \forall (x, t) \in D_T, \quad (4.10)$$

$$u_{xx}(x, t) = \frac{1}{2\sqrt{\pi t}} \int_{-\infty}^{\infty} u_0''(s) e^{-\frac{(x-s)^2}{4t}} ds$$

$$= \frac{1}{\sqrt{\pi}} \int_{-\infty}^{\infty} u_0''(x + 2\sqrt{t}w) e^{-w^2} dw; \quad \forall (x, t) \in D_T, \quad (4.11)$$

$$u_t(x, t) = \frac{1}{2\sqrt{\pi t}} \int_{-\infty}^{\infty} u_0''(s) e^{-\frac{(x-s)^2}{4t}} ds$$

$$= \frac{1}{\sqrt{\pi}} \int_{-\infty}^{\infty} u_0''(x + 2\sqrt{t}w) e^{-w^2} dw; \quad \forall (x, t) \in D_T. \quad (4.12)$$

We now have.

Theorem 4.2 (Global Existence and Uniqueness) *The problem (B-D-C) has exactly one solution on $\bar{D}_T$ for any $T > 0$. The solution is given by $u : \bar{D}_T \to \mathbb{R}$ as defined in (4.8) (and (4.9)).*

Proof Let $u : \bar{D}_T \to \mathbb{R}$ be as given in (4.8). It follows that u is continuous on $\bar{D}_T$ and that u_t, u_x and u_{xx} all exist and are continuous on D_T. Moreover, from (4.11) and (4.12) it follows that

$$u_t = u_{xx} \text{ on } D_T. \quad (4.13)$$

Also, via (4.8), we have

$$u(x, 0) = u_0(x); \quad \forall x \in \mathbb{R}. \quad (4.14)$$

Since $u_0 : \mathbb{R} \to \mathbb{R}$ is bounded, then there is a constant $M > 0$ such that

$$|u_0(x)| \le M; \quad \forall x \in \mathbb{R}. \quad (4.15)$$

It then follows from (4.9) that

$$|u(x, t)| \le M; \quad \forall (x, t) \in \bar{D}_T \quad (4.16)$$

and hence, $u(x, t)$ is uniformly bounded as $|x| \to \infty$ for $t \in [0, T]$.

Thus $u : \bar{D}_T \to \mathbb{R}$, as given in (4.8), provides a solution to (B-D-C) on $\bar{D}_T$ for any $T > 0$. That this is the only solution to (B-D-C) on $\bar{D}_T$ follows via Theorem 4.1. □

Remark 4.3 Since (B-D-C) has a (unique) solution on $\bar{D}_T$ for any $T > 0$, then (B-D-C) has a (unique) solution on $\bar{D}_\infty$. We say that (B-D-C) has a (unique) global solution on $\bar{D}_\infty$. ⌐

Theorem† **4.4** (Continuous Dependence) *Given $\epsilon > 0$, there exists $\delta > 0$ (depending only upon ϵ) such that for all $u_{10}, u_{20} \in BPC^2(\mathbb{R})$ such that*

$$\sup_{x \in \mathbb{R}} |u_{10}(x) - u_{20}(x)| < \delta,$$

then the corresponding solutions $u_1, u_2 : \bar{D}_\infty \to \mathbb{R}$ of (B-D-C) with initial data $u_0 = u_{10}$ and $u_0 = u_{20}$ respectively, satisfy

$$|u_1(x, t) - u_2(x, t)| < \epsilon; \quad \forall (x, t) \in \bar{D}_\infty.$$

Proof Let $u_1, u_2 : \bar{D}_\infty \to \mathbb{R}$ be the solutions to (B-D-C) as above, then via (4.9),

$$|u_1(x, t) - u_2(x, t)| \leq \frac{1}{\sqrt{\pi}} \int_{-\infty}^{\infty} \left| u_{10}(x + 2\sqrt{t}\lambda) - u_{20}(x + 2\sqrt{t}\lambda) \right| e^{-\lambda^2} d\lambda$$

$$\leq \sup_{x \in \mathbb{R}} |u_{10}(x) - u_{20}(x)|; \quad \forall (x, t) \in \bar{D}_\infty. \qquad (4.17)$$

Upon setting $\delta = \epsilon$, (4.17) yields

$$|u_1(x, t) - u_2(x, t)| < \epsilon; \quad \forall (x, t) \in \bar{D}_\infty,$$

as required. ☐

Remark 4.5 It is readily established from Theorem 4.4 that for every $u_0 \in BPC^2(\mathbb{R})$, the corresponding unique global solution $u : \bar{D}_\infty \to \mathbb{R}$ to (B-D-C) is Liapunov stable with respect to perturbations in the initial data $\delta u_0 \in BPC^2(\mathbb{R})$. ⌐

We can now state a classical result regarding (B-D-C),

Theorem† **4.6** *The problem (B-D-C) is uniformly globally well-posed on $BPC^2(\mathbb{R})$.*

Proof Observe that (P1) and (P2) are satisfied via Theorem 4.2. Similarly, (P3) is satisfied via Theorem 4.4. Moreover, δ depends only upon ϵ in Theorem 4.4. The result follows. ☐

We next examine some fundamental qualitative properties of the solution to (B-D-C) on $\bar{D}_\infty$, which we will require in the later chapters of the monograph. We have.

Theorem† 4.7 (Bounds and Derivative Estimates) *Let* $u : \bar{D}_T \to \mathbb{R}$ *be the unique solution to (B-D-C) on* $\bar{D}_\infty$ *and* M_0, M_0', M_0'', m_0, m_0' *and* m_0'' *be constants such that*

$$m_0 \leq u_0(x) \leq M_0, \ m_0' \leq u_0'(x) \leq M_0' \ and \ m_0'' \leq u_0''(x) \leq M_0'' \qquad (4.18)$$

for all $x \in \mathbb{R}$. *Then,*

$$m_0 \leq u(x, t) \leq M_0, \ m_0' \leq u_x(x, t) \leq M_0',$$
$$m_0'' \leq u_{xx}(x, t) \leq M_0'' \ and \ m_0'' \leq u_t(x, t) \leq M_0'' \qquad (4.19)$$

for all $(x, t) \in D_\infty$.

Proof Let $u : \bar{D}_\infty \to \mathbb{R}$ be the unique solution to (B-D-C). Then u is given by (4.8) and (4.9) on $\bar{D}_\infty$ whilst u_t, u_x and u_{xx} are given by (4.10), (4.11) and (4.12) on D_∞. The inequalities on D_∞ follow immediately. $\qquad \square$

We also note the following.

Remark 4.8 Let $u : \bar{D}_\infty \to \mathbb{R}$ be the unique solution to (B-D-C) on $\bar{D}_\infty$. It is straightforward to establish from (4.9) that when

$$u_0(x) \to \begin{cases} l^+ & \text{as } x \to +\infty \\ l^- & \text{as } x \to -\infty \end{cases} \qquad (4.20)$$

with $l^+, l^- \in \mathbb{R}$ constants. Then,

$$u(x, t) \to \begin{cases} l^+ & \text{as } x \to +\infty \\ l^- & \text{as } x \to -\infty \end{cases} \qquad (4.21)$$

uniformly for $t \in [0, T]$, for any $T > 0$. $\qquad \lrcorner$

We conclude this chapter with a result concerning an initial value problem for the inhomogeneous version of (B-D-C). We refer to this problem as (I-B-D-C), which differs from (B-D-C) only in that the diffusion equation in (i) is replaced by the inhomogeneous diffusion equation

$$u_t = u_{xx} + F(x, t), \quad \forall (x, t) \in D_T, \qquad (4.22)$$

where $F : \bar{D}_T \to \mathbb{R}$ is a bounded function which is continuous. We have the following fundamental result concerning (I-B-D-C) which will be of importance in later chapters. The proof is included for illustrative purposes; alternatively, see [59] (Section 6.4).

Theorem **4.9** *Let $u : \bar{D}_T \to \mathbb{R}$ be a solution to (I-B-D-C). Then,*

$$u(x,t) = \frac{1}{\sqrt{\pi}} \int_{-\infty}^{\infty} u_0 \left(x + 2\sqrt{t}\lambda \right) e^{-\lambda^2} d\lambda$$

$$+ \frac{1}{\sqrt{\pi}} \int_0^t \int_{-\infty}^{\infty} F \left(x + 2\sqrt{t - \tau}\lambda, \tau \right) e^{-\lambda^2} d\lambda d\tau$$

for all $(x, t) \in \bar{D}_T$.

Proof The proof will be based on the finite Laplace Transform. Let $u : \bar{D}_T \to \mathbb{R}$ be a solution to (I-B-D-C). Introduce $\hat{t} = t - \epsilon$ with $0 < \epsilon < T$, and write

$$u(x,t) = \hat{u}(x,\hat{t}), \quad F(x,t) = \hat{F}(x,\hat{t}) \tag{4.23}$$

for all $(x, \hat{t}) \in \mathbb{R} \times [0, T - \epsilon]$. Now, introduce $C^+ = \{k \in \mathbb{C} : \text{Re}(k) \geq 1\}$ and define $\bar{u} : \mathbb{R} \times C^+ \to \mathbb{C}$ as

$$\bar{u}(x,k) = \int_0^{T-\epsilon} \hat{u}(x,\hat{t}) e^{-k\hat{t}} d\hat{t} \tag{4.24}$$

for all $(x, k) \in \mathbb{R} \times C^+$. Similarly, define $\bar{F} : \mathbb{R} \times C^+ \to \mathbb{C}$ as

$$\bar{F}(x,k) = \int_0^{T-\epsilon} \hat{F}(x,\hat{t}) e^{-k\hat{t}} d\hat{t} \tag{4.25}$$

for all $(x, k) \in \mathbb{R} \times C^+$. It is readily established from (4.24) and regularity on $\hat{u}$, that $\bar{u}$ is continuous and bounded on $\mathbb{R} \times C^+$, whilst $\bar{u}_x$, $\bar{u}_{xx}$ and $\bar{u}_k$ all exist and are continuous on $\mathbb{R} \times C^+$. In particular then $\bar{u}(x, \cdot)$ is analytic on C^+ for each $x \in \mathbb{R}$, with

$$|\bar{u}(x,k)| \leq \frac{M}{|k|} \quad \text{as } |k| \to \infty \text{ on } C^+ \tag{4.26}$$

where

$$M = \sup_{(x,t) \in \bar{D}_T} |u(x,t)|.$$

Similarly, $\bar{F}$ is continuous and bounded on $\mathbb{R} \times C^+$ with $\bar{F}_k$ continuous on $\mathbb{R} \times C^+$, and so $\bar{F}(x, \cdot)$ is analytic on C^+ for each $x \in \mathbb{R}$. Moreover,

$$|\bar{F}(x,k)| \leq \frac{N}{|k|} \quad \text{as } |k| \to \infty \text{ on } C^+ \tag{4.27}$$

with

$$N = \sup_{(x,t) \in \bar{D}_T} |F(x,t)|.$$

We also have the derivative formulae,

$$\bar{u}_x(x, k) = \int_0^{T-\epsilon} \hat{u}_x(x, \hat{t}) e^{-k\hat{t}} d\hat{t}$$

$$\bar{u}_{xx}(x, k) = \int_0^{T-\epsilon} \hat{u}_{xx}(x, \hat{t}) e^{-k\hat{t}} d\hat{t} \qquad (4.28)$$

$$k\bar{u}(x, k) = -\hat{u}(x, T - \epsilon) e^{-k(T-\epsilon)} + \hat{u}(x, 0) + \int_0^{T-\epsilon} \hat{u}_{\hat{t}}(x, \hat{t}) e^{-k\hat{t}} d\hat{t}$$

for all $(x, k) \in \mathbb{R} \times C^+$ (which follows via standard uniform convergence results). Now,

$$\hat{u}_{\hat{t}}(x, \hat{t}) - \hat{u}_{xx}(x, \hat{t}) = \hat{F}(x, \hat{t}) \qquad (4.29)$$

for all $(x, \hat{t}) \in \mathbb{R} \times [0, T - \epsilon]$. Next multiply (4.29) by $e^{-k\hat{t}}$, with $(k, \hat{t}) \in C^+ \times [0, T - \epsilon]$, and integrate with respect $\hat{t}$ from $\hat{t} = 0$ to $\hat{t} = T - \epsilon$, from which we obtain, via (4.24)-(4.28),

$$\bar{u}_{xx}(x, k) - k\bar{u}(x, k) = -\bar{F}(x, k) + \hat{u}(x, T - \epsilon) e^{-k(T-\epsilon)} - \hat{u}(x, 0) \qquad (4.30)$$

for all $(x, k) \in \mathbb{R} \times C^+$, whilst from (4.24),

$$|\bar{u}(x, k)| \leq MT \qquad (4.31)$$

for all $(x, k) \in \mathbb{R} \times C^+$. It follows from (4.30) and (4.31) that,

$$\bar{u}(x, k) = \int_{-\infty}^{\infty} G(x - \bar{x}, k) \left(\hat{u}(\bar{x}, 0) - \hat{u}(\bar{x}, T - \epsilon) e^{-k(T-\epsilon)} + \bar{F}(\bar{x}, k) \right) d\bar{x} \qquad (4.32)$$

for $(x, k) \in \mathbb{R} \times C^+$, with $G : \mathbb{R} \times C^+ \to \mathbb{C}$ given by,

$$G(\lambda, k) = \frac{1}{2k^{1/2}} e^{-k^{1/2}|\lambda|} \qquad (4.33)$$

for all $(\lambda, k) \in \mathbb{R} \times C^+$, and $k^{1/2}$ has a branch-cut along the negative real k-axis, with $\text{Re}(k^{1/2}) > 0$ in the cut-k-plane (preserving analyticity for $k \in C^+$). The Bromwich Inversion Theorem (see [18], Chapter 3), via (4.24) and (4.33), gives,

$$\hat{u}(x, \hat{t}) = \int_{-\infty}^{\infty} \hat{u}(\bar{x}, 0) \left(\int_{1-i\infty}^{1+i\infty} G(x - \bar{x}, k) e^{k\hat{t}} dk \right) d\bar{x}$$

$$+ \int_{-\infty}^{\infty} \left(\int_{1-i\infty}^{1+i\infty} \bar{F}(\bar{x}, k) G(x - \bar{x}, k) e^{k\hat{t}} dk \right) d\bar{x}$$

$$- \int_{-\infty}^{\infty} \hat{u}(\bar{x}, T - \epsilon) \left(\int_{1-i\infty}^{1+i\infty} G(x - \bar{x}, k) e^{k(\hat{t}-(T-\epsilon))} dk \right) d\bar{x}$$

$$(4.34)$$

for all $(x, \hat{t}) \in \mathbb{R} \times (0, T - \epsilon)$. It follows from the Cauchy Residue Theorem, with (4.27), after deforming the contour onto a large semi-circle in the right hand of the k-plane, that

$$\int_{1-i\infty}^{1+i\infty} G(x - \bar{x}, k)e^{k(\hat{t}-(T-\epsilon))}dk = 0 \qquad (4.35)$$

for all $(x, \bar{x}, \hat{t}) \in \mathbb{R}^2 \times [0, T-\epsilon)$. Also, standard Bromwich inversion formulae give,

$$\int_{1-i\infty}^{1+i\infty} G(x - \bar{x}, k)e^{k\hat{t}}dk = \frac{1}{2\sqrt{\pi}\hat{t}^{1/2}}e^{-\frac{(x-\bar{x})^2}{4\hat{t}}} \qquad (4.36)$$

for all $(x, \bar{x}, \hat{t}) \in \mathbb{R}^2 \times (0, T-\epsilon]$. The Laplace Convolution Theorem (see [18], Chapter 2) finally gives,

$$\int_{1-i\infty}^{1+i\infty} \bar{F}(\bar{x}, k)G(x-\bar{x}, k)e^{k\hat{t}}dk = \frac{1}{2\sqrt{\pi}}\int_0^{\hat{t}^-} \frac{\hat{F}(\bar{x}, \hat{\tau})}{(\hat{t} - \hat{\tau})^{1/2}}e^{-\frac{(x-\bar{x})^2}{4(\hat{t}-\hat{\tau})}}d\hat{\tau} \qquad (4.37)$$

for all $(x, \bar{x}, \hat{t}) \in \mathbb{R}^2 \times (0, T - \epsilon]$. On substitution from (4.35)–(4.37) into (4.34) we obtain,

$$\hat{u}(x, \hat{t}) = \frac{1}{2\sqrt{\pi}\hat{t}^{1/2}} \int_{-\infty}^{\infty} \hat{u}(\bar{x}, 0)e^{-\frac{(x-\bar{x})^2}{4\hat{t}}}d\bar{x}$$

$$+ \frac{1}{2\sqrt{\pi}}\int_0^{\hat{t}^-}\int_{-\infty}^{\infty} \frac{\hat{F}(\bar{x}, \hat{\tau})}{(\hat{t} - \hat{\tau})^{1/2}}e^{-\frac{(x-\bar{x})^2}{4(\hat{t}-\hat{\tau})}}d\bar{x}d\hat{\tau} \qquad (4.38)$$

for all $(x, \hat{t}) \in \mathbb{R} \times (0, T-\epsilon)$. Using (4.23) together with a simple substitution enables (4.38) to be written as,

$$u(x, t) = \frac{1}{\sqrt{\pi}} \int_{-\infty}^{\infty} u(x + 2\sqrt{t - \epsilon}s, \epsilon)e^{-s^2}ds$$

$$+ \frac{1}{\sqrt{\pi}}\int_{\epsilon}^{t}\int_{-\infty}^{\infty} F(x + 2\sqrt{t - \tau}\lambda, \tau)e^{-\lambda^2}d\lambda d\tau \qquad (4.39)$$

for all $(x, t) \in \mathbb{R} \times (\epsilon, T)$ whilst continuity of u and F on $\bar{D}_T$ allows (4.39) to be extended to $(x, t) \in \mathbb{R} \times [\epsilon, T]$. With $F, u \in C(\bar{D}_T)$, together with uniform convergence of the improper integrals in (4.39), the limit as $\epsilon \to 0$ can be taken in (4.39) to obtain,

$$u(x, t) = \frac{1}{\sqrt{\pi}} \int_{-\infty}^{\infty} u_0(x + 2\sqrt{t}\lambda)e^{-\lambda^2}d\lambda$$

$$+ \frac{1}{\sqrt{\pi}}\int_0^{t}\int_{-\infty}^{\infty} F(x + 2\sqrt{t - \tau}\lambda, \tau)e^{-\lambda^2}d\lambda d\tau$$

for all $(x, t) \in \bar{D}_T$, as required. $\square$

5

Convolution Functions, Function Spaces, Integral Equations and Equivalence Lemmas

This chapter contains a result, which relates solutions of (B-R-D-C) with $f \in H_\alpha$ to continuous, bounded solutions of an implicit integral equation. From this result, we obtain derivative bounds of Schauder type on solutions of (B-R-D-C) with $f \in H_\alpha$. It should be noted that results marked with the subscript † correspond to results in [21] (Chapter 1, Sections 6 and 7) for which sketched proofs are given (unless otherwise stated).

5.1 Convolution Functions

Let $F : \bar{D}_T \to \mathbb{R}$ be continuous and bounded. Thus, there exists a constant $M_T > 0$ such that

$$|F(x, t)| \le M_T; \quad \forall (x, t) \in \bar{D}_T. \tag{5.1}$$

Define the convolution function $\phi : \bar{D}_T \to \mathbb{R}$ as

$$\phi(x, t) = \frac{1}{\sqrt{\pi}} \int_0^t \int_{-\infty}^{\infty} F\left(x + 2\sqrt{t - \tau}\, w, \tau\right) e^{-w^2} dw d\tau; \quad \forall (x, t) \in \bar{D}_T. \tag{5.2}$$

It is readily established that ϕ is well-defined on $\bar{D}_T$. Also,

$$\phi(x, 0) = 0; \quad \forall x \in \mathbb{R}. \tag{5.3}$$

In addition, $\phi : \bar{D}_T \to \mathbb{R}$ is continuous and bounded with

$$|\phi(x, t)| \le M_T T; \quad \forall (x, t) \in \bar{D}_T. \tag{5.4}$$

We next define, on $\bar{D}_T^\delta$ ($0 < \delta < T$), the sequence of functions $\phi_n : \bar{D}_T^\delta \to \mathbb{R}$ for $n = N_\delta, N_\delta + 1, \ldots$, with $N_\delta = [\delta^{-1}] + 1$, as

$$\phi_n(x,t) = \frac{1}{\sqrt{\pi}} \int_0^{t-1/n} \int_{-\infty}^{\infty} F\left(x + 2\sqrt{t-\tau}\,w, \tau\right) e^{-w^2} dw d\tau; \quad \forall (x,t) \in \bar{D}_T^\delta. \tag{5.5}$$

The function ϕ_n ($n = N_\delta, N_\delta + 1, \ldots$) has the following properties:

(a) ϕ_n is continuous on $\bar{D}_T^\delta$,
(b) ϕ_n is bounded on $\bar{D}_T^\delta$, with $|\phi_n(x,t)| \le M_T T \;\forall (x,t) \in \bar{D}_T^\delta$, and
(c) $\phi_n(x,t) \to \phi(x,t)$ as $n \to \infty$ uniformly $\forall (x,t) \in \bar{D}_T^\delta$.

We now observe that by a simple substitution $\left(s = x + 2\sqrt{t-\tau}\,w\right)$, we may write

$$\phi_n(x,t) = \frac{1}{2\sqrt{\pi}} \int_0^{t-1/n} \int_{-\infty}^{\infty} \frac{F(s,\tau)}{(t-\tau)^{1/2}} e^{-\frac{(s-x)^2}{4(t-\tau)}} ds d\tau; \quad \forall (x,t) \in \bar{D}_T^\delta. \tag{5.6}$$

It follows from (5.6), via standard results on uniform convergence of integrals [6] that

$$\phi_{nx}, \;\; \phi_{nxx} \;\; \text{and} \;\; \phi_{nt} \tag{5.7}$$

all exist and are continuous on $\bar{D}_T^\delta$, with

$$\phi_{nx}(x,t) = \frac{1}{\sqrt{\pi}} \int_0^{t-1/n} \int_{-\infty}^{\infty} \frac{F\left(x + 2\sqrt{t-\tau}\,w, \tau\right)}{(t-\tau)^{1/2}} w e^{-w^2} dw d\tau; \quad \forall (x,t) \in \bar{D}_T^\delta, \tag{5.8}$$

$$\phi_{nxx}(x,t) = \frac{1}{\sqrt{\pi}} \int_0^{t-1/n} \int_{-\infty}^{\infty} \frac{F\left(x + 2\sqrt{t-\tau}\,w, \tau\right)}{(t-\tau)} (w^2 - 1/2) e^{-w^2} dw d\tau;$$
$$\forall (x,t) \in \bar{D}_T^\delta, \tag{5.9}$$

$$\phi_{nt}(x,t) = \frac{1}{\sqrt{\pi}} \int_0^{t-1/n} \int_{-\infty}^{\infty} \frac{F\left(x + 2\sqrt{t-\tau}\,w, \tau\right)}{(t-\tau)} (w^2 - 1/2) e^{-w^2} dw d\tau$$
$$+ \frac{1}{\sqrt{\pi}} \int_{-\infty}^{\infty} F\left(x + 2\sqrt{1/n}\,w, t - 1/n\right) e^{-w^2} dw; \quad \forall (x,t) \in \bar{D}_T^\delta. \tag{5.10}$$

We observe from (5.8) that

$$|\phi_{nx}(x,t)| \le \frac{M_T}{\sqrt{\pi}} \int_0^{t-1/n} \int_{-\infty}^{\infty} \frac{1}{(t-\tau)^{1/2}} |w| e^{-w^2} dw d\tau$$
$$= \frac{M_T}{\sqrt{\pi}} \left[-2(t-\tau)^{1/2} \right]_0^{t-1/n}$$
$$= \frac{2M_T}{\sqrt{\pi}} \left(t^{1/2} - (1/n)^{1/2} \right)$$

$$\leq \frac{2M_T}{\sqrt{\pi}} \left(T^{1/2} + 1\right); \quad \forall (x,t) \in \bar{D}_T^\delta, \tag{5.11}$$

and so ϕ_{nx} is bounded on $\bar{D}_T^\delta$, uniformly in n. We now have:

Theorem† 5.1 $\phi : \bar{D}_T \to \mathbb{R}$ *in (5.2) is such that ϕ_x exists and is continuous and bounded on D_T, with*

$$|\phi_x(x,t)| \leq \frac{2M_T}{\sqrt{\pi}} \left(T^{1/2} + 1\right); \quad \forall (x,t) \in D_T.$$

Proof First we recall that ϕ_n and ϕ_{nx} are continuous and bounded on $\bar{D}_T^\delta$ and that $\phi_n \to \phi$ as $n \to \infty$ uniformly on $\bar{D}_T^\delta$. Now let $n \geq m \geq N_\delta$ and $(x,t) \in \bar{D}_T^\delta$, then

$$|\phi_{nx}(x,t) - \phi_{mx}(x,t)|$$

$$= \left| \frac{1}{\sqrt{\pi}} \int_{t-1/m}^{t-1/n} \int_{-\infty}^{\infty} \frac{F\left(x + 2\sqrt{t-\tau}\, w, \tau\right)}{(t-\tau)^{1/2}} w e^{-w^2} dw d\tau \right|$$

$$\leq \frac{M_T}{\sqrt{\pi}} \int_{t-1/m}^{t-1/n} \int_{-\infty}^{\infty} \frac{1}{(t-\tau)^{1/2}} |w| e^{-w^2} dw d\tau$$

$$= \frac{2M_T}{\sqrt{\pi}} \left(\left(\frac{1}{m}\right)^{1/2} - \left(\frac{1}{n}\right)^{1/2} \right)$$

$$\leq \frac{2M_T}{\sqrt{\pi}} (1/m + 1/n); \quad \forall (x,t) \in \bar{D}_T^\delta.$$

It follows that $\{\phi_{nx}\}$ is uniformly convergent on $\bar{D}_T^\delta$ as $n \to \infty$, via the Cauchy condition [66], and moreover via Theorem 7.17 in [66], that ϕ_x exists, is continuous and bounded on $\bar{D}_T^\delta$, with

$$\phi_{nx} \to \phi_x \text{ as } n \to \infty \text{ uniformly on } \bar{D}_T^\delta. \tag{5.12}$$

Now, given any $\epsilon > 0$, there exists $r \geq N_\delta$ such that

$$|\phi_x(x,t) - \phi_{rx}(x,t)| < \epsilon; \quad \forall (x,t) \in \bar{D}_T^\delta, \tag{5.13}$$

via (5.12). Hence, using (5.11),

$$|\phi_x(x,t)| < \epsilon + \frac{2M_T}{\sqrt{\pi}} \left(T^{1/2} + 1\right); \quad \forall (x,t) \in \bar{D}_T^\delta. \tag{5.14}$$

However, (5.14) holds for any $\epsilon > 0$, and so

$$|\phi_x(x,t)| \leq \frac{2M_T}{\sqrt{\pi}} \left(T^{1/2} + 1\right); \quad \forall (x,t) \in \bar{D}_T^\delta. \tag{5.15}$$

Now all of the above holds for any fixed $0 < \delta < T$, and so it follows that ϕ_x exists, is continuous and bounded on D_T, with

$$|\phi_x(x,t)| \le \frac{2M_T}{\sqrt{\pi}} \left(T^{1/2} + 1\right); \quad \forall (x,t) \in D_T. \qquad (5.16)$$

The proof is complete.   □

We now restrict $F : \bar{D}_T \to \mathbb{R}$ to satisfy the additional condition:

(H) $F : \bar{D}_T \to \mathbb{R}$ is continuous, bounded and uniformly Hölder continuous of degree $0 < \alpha \le 1$ with respect to $x \in \mathbb{R}$, uniformly for $t \in [0, T]$. That is, there exists a constant $k_T > 0$ (independent of $t \in [0, T]$) such that

$$|F(y,t) - F(x,t)| \le k_T |y - x|^\alpha; \quad \forall (y,t), (x,t) \in \bar{D}_T.$$

⌐

We now observe that, for $(x,t) \in \bar{D}_T^\delta$,

$$
\begin{aligned}
|\phi_{nxx}(x,t)| &\le \left| \frac{1}{\sqrt{\pi}} \int_0^{t-1/n} \int_{-\infty}^{\infty} \frac{F\left(x + 2\sqrt{t-\tau}\, w, \tau\right) - F(x,\tau)}{(t-\tau)} \right. \\
&\qquad \left. (w^2 - 1/2) e^{-w^2} dw\, d\tau \right| \\
&\quad + \left| \frac{1}{\sqrt{\pi}} \int_0^{t-1/n} \int_{-\infty}^{\infty} \frac{F(x,\tau)}{(t-\tau)} (w^2 - 1/2) e^{-w^2} dw\, d\tau \right| \quad (= 0) \\
&\le \frac{k_T}{\sqrt{\pi}} \int_0^{t-1/n} \int_{-\infty}^{\infty} \frac{2^\alpha}{(t-\tau)^{1-\alpha/2}} |w|^\alpha |w^2 - 1/2| e^{-w^2} dw\, d\tau \\
&= \frac{2^\alpha k_T}{\sqrt{\pi}} I_\alpha \left[\frac{-2}{\alpha} (t-\tau)^{\alpha/2} \right]_0^{t-1/n} \\
&= \frac{2^{\alpha+1} k_T}{\alpha\sqrt{\pi}} I_\alpha \left(t^{\alpha/2} - (1/n)^{\alpha/2} \right) \\
&\le \frac{2^{\alpha+1} k_T}{\alpha\sqrt{\pi}} I_\alpha \left(1 + T^{\alpha/2} \right); \quad \forall (x,t) \in \bar{D}_T^\delta, \qquad (5.17)
\end{aligned}
$$

with

$$I_\alpha = \int_{-\infty}^{\infty} |w|^\alpha |w^2 - 1/2| e^{-w^2} dw > 0. \qquad (5.18)$$

Similarly,

$$|\phi_{nt}(x,t)| \le \frac{2^{\alpha+1} k_T}{\alpha\sqrt{\pi}} I_\alpha \left(1 + T^{\alpha/2} \right) + M_T; \quad \forall (x,t) \in \bar{D}_T^\delta. \qquad (5.19)$$

Thus, under condition (H), both ϕ_{nt} and ϕ_{nxx} are continuous and bounded (uniformly in n) on $\bar{D}_T^\delta$, for each $n = N_\delta$, $N_\delta + 1$, ...

We next observe the following:

(I) With $n \geq m \geq N_\delta$ and $(x, t) \in \bar{D}_T^\delta$,

$$\left| \int_{t-1/m}^{t-1/n} \int_{-\infty}^{\infty} \frac{F\left(x + 2\sqrt{t-\tau}\, w, \tau\right)}{(t-\tau)} (w^2 - 1/2) e^{-w^2} dw\, d\tau \right|$$

$$\leq \int_{t-1/m}^{t-1/n} \int_{-\infty}^{\infty} \frac{\left| F\left(x + 2\sqrt{t-\tau}\, w, \tau\right) - F(x, \tau) \right|}{(t-\tau)} |w^2 - 1/2| e^{-w^2} dw\, d\tau$$

$$+ \left| \int_{t-1/m}^{t-1/n} \int_{-\infty}^{\infty} \frac{F(x, \tau)}{(t-\tau)} (w^2 - 1/2) e^{-w^2} dw\, d\tau \right| \quad (= 0)$$

$$\leq k_T \int_{t-1/m}^{t-1/n} \int_{-\infty}^{\infty} \frac{2^\alpha}{(t-\tau)^{1-\alpha/2}} |w|^\alpha |w^2 - 1/2| e^{-w^2} dw\, d\tau$$

$$= k_T 2^\alpha I_\alpha \left[-\frac{2}{\alpha} (t-\tau)^{\alpha/2} \right]_{t-1/m}^{t-1/n}$$

$$= \frac{2^{\alpha+1} k_T I_\alpha}{\alpha} \left((1/m)^{\alpha/2} - (1/n)^{\alpha/2} \right); \quad \forall (x, t) \in \bar{D}_T^\delta. \tag{5.20}$$

(II) Let $n \geq N_\delta$. Given any $\epsilon > 0$, there exists $\sigma > 0$ (depending upon ϵ, δ, X, T) such that for all $(x_0, t_0), (x_1, t_1) \in \bar{D}_T^{\delta - 1/N_\delta, X}$ with

$$|(x_1 - x_0, t_1 - t_0)| < \sigma,$$

then

$$|F(x_1, t_1) - F(x_0, t_0)| < \epsilon/2,$$

since F is continuous and therefore uniformly continuous on $\bar{D}_T^{\delta - 1/N_\delta, X}$. Now let $(x, t) \in \bar{D}_T^{\delta, X}$, then

$$\left| \frac{1}{\sqrt{\pi}} \int_{-\infty}^{\infty} F\left(x + 2\sqrt{1/n}\, w, t - 1/n\right) e^{-w^2} dw - F(x, t) \right|$$

$$\leq \frac{1}{\sqrt{\pi}} \int_{-\infty}^{\infty} |F\left(x + 2\sqrt{1/n}\, w, t - 1/n\right)$$

$$- F(x, t - 1/n)| e^{-w^2} dw + |F(x, t - 1/n) - F(x, t)|$$

$$\leq \frac{k_T}{\sqrt{\pi}} \int_{-\infty}^{\infty} \frac{2^\alpha}{n^{\alpha/2}} |w|^\alpha e^{-w^2} dw + |F(x, t - 1/n) - F(x, t)|$$

$$\leq \frac{2^\alpha k_T J_\alpha}{\sqrt{\pi}} \frac{1}{n^{\alpha/2}} + |F(x, t - 1/n) - F(x, t)| \tag{5.21}$$

where

$$J_\alpha = \int_{-\infty}^{\infty} |w|^\alpha e^{-w^2} dw > 0. \tag{5.22}$$

Now since $(x, t) \in \bar{D}_T^{\delta, X}$ and $n \geq N_\delta$, then

$$(x, t - 1/n), (x, t) \in \bar{D}_T^{\delta - 1/N_\delta, X}.$$

Take $n > 1/\sigma + 1$, and so

$$|(x, t - 1/n) - (x, t)| < \sigma; \quad \forall (x, t) \in \bar{D}_T^{\delta, X},$$

and so

$$|F(x, t - 1/n) - F(x, t)| < \epsilon/2; \quad \forall (x, t) \in \bar{D}_T^{\delta, X}. \tag{5.23}$$

Therefore, given any $\epsilon > 0$, then for all

$$n > \max \left\{ 1/\sigma + 1, \left(\frac{2^{\alpha+1} k_T J_\alpha}{\sqrt{\pi} \epsilon} \right)^{2/\alpha} + 1 \right\},$$

we have

$$\left| \frac{1}{\sqrt{\pi}} \int_{-\infty}^{\infty} F\left(x + 2\sqrt{1/n}\, w, t - 1/n\right) e^{-w^2} dw - F(x, t) \right|$$

$$< \epsilon/2 + \epsilon/2 = \epsilon; \quad \forall (x, t) \in \bar{D}_T^{\delta, X}.$$

Thus,

$$\frac{1}{\sqrt{\pi}} \int_{-\infty}^{\infty} F\left(x + 2\sqrt{1/n}\, w, t - 1/n\right) e^{-w^2} dw \to F(x, t)$$

$$\text{as } n \to \infty \text{ uniformly on } \bar{D}_T^{\delta, X} \text{ (any } \delta, X > 0). \tag{5.24}$$

We now have:

Theorem† 5.2 *The function* $\phi : \bar{D}_T \to \mathbb{R}$ *in* (5.2) *is such that* ϕ_t *and* ϕ_{xx} *exist, are continuous and bounded on* D_T, *with*

$$|\phi_{xx}(x, t)| \leq \frac{2^{\alpha+1} k_T I_\alpha}{\alpha \sqrt{\pi}} (1 + T^{\alpha/2}), \tag{5.25}$$

$$|\phi_t(x, t)| \leq \frac{2^{\alpha+1} k_T I_\alpha}{\alpha \sqrt{\pi}} (1 + T^{\alpha/2}) + M_T. \tag{5.26}$$

Moreover,

$$\phi_t(x, t) = \phi_{xx}(x, t) + F(x, t); \quad \forall (x, t) \in D_T. \tag{5.27}$$

Proof First we recall that ϕ_n and ϕ_{nx} are continuous and bounded uniformly in n on $\bar{D}_T^\delta$ and that $\phi_n \to \phi$ and $\phi_{nx} \to \phi_x$ as $n \to \infty$ uniformly on $\bar{D}_T^\delta$. Moreover, ϕ_{nxx} is continuous and bounded uniformly in n on $\bar{D}_T^\delta$. Now let $n \geq m \geq N_\delta$ and $(x,t) \in \bar{D}_T^\delta$, then it follows from (5.20) that

$$|\phi_{nxx}(x,t) - \phi_{mxx}(x,t)| \leq \frac{2^{\alpha+1} k_T I_\alpha}{\alpha \sqrt{\pi}} \left((1/m)^{\alpha/2} + (1/n)^{\alpha/2} \right); \quad \forall (x,t) \in \bar{D}_T^\delta.$$

It follows that $\{\phi_{nxx}\}$ is uniformly convergent on $\bar{D}_T^\delta$ as $n \to \infty$, via the Cauchy condition [66], and moreover, via Theorem 7.17 in [66], that ϕ_{xx} exists, is continuous and is bounded on $\bar{D}_T^\delta$, with

$$\phi_{nxx} \to \phi_{xx} \text{ as } n \to \infty \text{ uniformly on } \bar{D}_T^\delta. \tag{5.28}$$

It follows from (5.28) and (5.17) that

$$|\phi_{xx}(x,t)| \leq \frac{2^{\alpha+1} k_T I_\alpha}{\alpha \sqrt{\pi}} \left(1 + T^{\alpha/2} \right); \quad \forall (x,t) \in \bar{D}_T^\delta. \tag{5.29}$$

Again recall that ϕ_n and ϕ_{nt} are continuous and bounded uniformly in n on $\bar{D}_T^\delta$ and $\phi_n \to \phi$ as $n \to \infty$ uniformly on $\bar{D}_T^\delta$. It now follows from (5.10) together with (I) and (II) that $\{\phi_{nt}\}$ is uniformly convergent on $\bar{D}_T^{\delta,X}$ (any $X > 0$) as $n \to \infty$, and so, moreover, that ϕ_t exists, and is continuous on $\bar{D}_T^{\delta,X}$. Now, given any $\epsilon > 0$ there exists $r \geq N_\delta$ such that

$$|\phi_t(x,t) - \phi_{rt}(x,t)| < \epsilon; \quad \forall (x,t) \in \bar{D}_T^{\delta,X}. \tag{5.30}$$

Hence using (5.19),

$$|\phi_t(x,t)| < \epsilon + \frac{2^{\alpha+1} k_T I_\alpha}{\alpha \sqrt{\pi}} \left(1 + T^{\alpha/2} \right) + M_T; \quad \forall (x,t) \in \bar{D}_T^{\delta,X}. \tag{5.31}$$

However, (5.31) holds for any $\epsilon > 0$, and so

$$|\phi_t(x,t)| \leq \frac{2^{\alpha+1} k_T I_\alpha}{\alpha \sqrt{\pi}} \left(1 + T^{\alpha/2} \right) + M_T; \quad \forall (x,t) \in \bar{D}_T^{\delta,X}. \tag{5.32}$$

Now, all of the above holds for any $X > 0$. Thus, ϕ_t exists, is continuous and bounded on $\bar{D}_T^\delta$, with

$$|\phi_t(x,t)| \leq \frac{2^{\alpha+1} k_T I_\alpha}{\alpha \sqrt{\pi}} \left(1 + T^{\alpha/2} \right) + M_T; \quad \forall (x,t) \in \bar{D}_T^\delta. \tag{5.33}$$

We next observe that since all of the above holds for all $0 < \delta < T$, then ϕ_t and ϕ_{xx} exist and are continuous on D_T whilst (5.29) and (5.33) establish that ϕ_t and ϕ_{xx} are bounded on D_T with both (5.29) and (5.33) continuing to hold on D_T.

Finally, to obtain (5.27), let $(x, t) \in \bar{D}_T^{\delta, X}$, then it follows from (5.9), (5.10), (I) and (II) that

$$(\phi_{nt}(x, t) - \phi_{nxx}(x, t)) \to F(x, t) \text{ as } n \to \infty \text{ uniformly on } \bar{D}_T^{\delta, X}. \quad (5.34)$$

Also from (5.25) and (5.26), we have

$$(\phi_{nt}(x, t) - \phi_{nxx}(x, t)) \to \phi_t(x, t) - \phi_{xx}(x, t) \text{ as } n \to \infty \text{ uniformly on } \bar{D}_T^{\delta, X}. \quad (5.35)$$

Uniqueness of limits, together with (5.34) and (5.35), then gives

$$\phi_t(x, t) - \phi_{xx}(x, t) = F(x, t); \quad \forall(x, t) \in \bar{D}_T^{\delta, X}. \quad (5.36)$$

However, (5.36) holds for any $X > 0$ and $0 < \delta < T$ and so continues to hold on D_T. The proof is complete. $\qquad \square$

5.2 Function Spaces

Associated with the (B-R-D-C), we introduce the sets of functions,

$$B_A^T = \{u : \bar{D}_T \to \mathbb{R} : u \text{ is continuous and bounded on } \bar{D}_T\} \quad (5.37)$$

$$B_B = \{v : \mathbb{R} \to \mathbb{R} : v \text{ is continuous and bounded on } \mathbb{R}\}. \quad (5.38)$$

Remark 5.3

(i) It follows immediately from (5.37) and (5.38) that when $u(\cdot, \cdot) \in B_A^T$ then $u(\cdot, t) \in B_B$ for each $t \in [0, T]$.
(ii) Whenever $u : \bar{D}_T \to \mathbb{R}$ is a solution to (B-R-D-C), then $u \in B_A^T$, via Definition 2.1.
(iii) Both B_A^T and B_B form linear spaces over $\mathbb{R}$ under the usual definitions of addition and scalar multiplication of functions.
(iv) The notation for B_A^T and B_B has been adopted from [56] for brevity. However, note that these could be equivalently denoted as $B_A^T = C_B(\bar{D}_T)$ and $B_B = C_B(\mathbb{R})$, where for a set X, $C_B(X) := C(X) \cap L^\infty(X)$. $\quad \lrcorner$

We next introduce the *norms* defined on B_A^T and B_B, namely,

$$\|u\|_A = \sup_{(x,t)\in\bar{D}_T} |u(x, t)|; \quad \forall u \in B_A^T, \quad (5.39)$$

$$\|v\|_B = \sup_{x\in\mathbb{R}} |v(x)|; \quad \forall v \in B_B. \quad (5.40)$$

We observe from (5.39) and (5.40) that whenever $u \in B_A^T$, then

$$||u(\cdot, t)||_B \leq ||u||_A; \quad \forall t \in [0, T]. \tag{5.41}$$

Remark 5.4 (Completeness) Both B_A^T and B_B (with usual addition and scalar multiplication over $\mathbb{R}$) are Banach Spaces (that is they are complete with respect to $|| \cdot ||_A$ and $|| \cdot ||_B$ respectively). ⌐

The following elementary lemma will be useful in later chapters.

Lemma† 5.5 *Let* $u \in B_A^T$. *Then* $H : [0, T] \to \mathbb{R}^+ \cup \{0\}$, *defined by*

$$H(t) = ||u(\cdot, t)||_B; \quad \forall t \in [0, T],$$

is such that $H \in L^1([0, T])$.

Proof First, observe that $H : [0, T] \to \mathbb{R}^+ \cup \{0\}$ is well-defined, via Remark 5.3 (i). Also, $H : [0, T] \to \mathbb{R}^+ \cup \{0\}$ is bounded, with

$$0 \leq H(t) \leq ||u||_A, \tag{5.42}$$

via (5.41). Next introduce the sequence of functions $\{H_n : [0, T] \to \mathbb{R}^+ \cup \{0\}\}_{n \in \mathbb{N}}$ such that

$$H_n(t) = \sup_{x \in [-n, n]} |u(x, t)|; \quad \forall t \in [0, T]. \tag{5.43}$$

Since $u \in B_A^T$, then it follows from (5.43) that $H_n \in C([0, T]) \subset L^1([0, T])$, with

$$0 \leq H_n(t) \leq ||u||_A; \quad \forall t \in [0, T]. \tag{5.44}$$

In addition,

$$0 \leq H_1(t) \leq H_2(t) \leq \cdots \leq H_n(t) \leq \cdots \leq ||u||_A; \quad \forall t \in [0, T]. \tag{5.45}$$

Moreover, it follows from (5.45) that

$$H_n(t) \to H(t) \text{ as } n \to \infty \tag{5.46}$$

for each $t \in [0, T]$. It is an immediate consequence of (5.44), (5.45), (5.46) and the monotone convergence theorem ([74], Theorem 2, p96) that $H \in L^1([0, T])$. □

Next, a standard generalisation of Gronwall's inequality [24], which will be useful in later chapters, can be established.

Proposition† **5.6** (Generalised Gronwall's Inequality) *Let* $\phi : [0, T] \to \mathbb{R}$ *be such that* $\phi \in L^1([0, T])$ *and* $\phi(t) \geq 0$ *for all* $t \in [0, T]$. *Suppose that*

$$\phi(t) \leq a + bt + k \int_0^t \phi(s)ds; \quad \forall t \in [0, T]$$

with $a \geq 0$, $b \geq 0$, $k > 0$ *constants. Then,*

$$\phi(t) \leq (a + bt)e^{kt}; \quad \forall t \in [0, T].$$

Proof It follows from the first inequality above that

$$\phi(t)e^{-kt} + (-k)e^{-kt} \int_0^t \phi(s)ds \leq (a + bt)e^{-kt}; \quad \forall t \in [0, T]. \tag{5.47}$$

Since $\phi \in L^1([0, T])$, it then follows from [74] (Proposition 2, p. 103), that, after an integration, (5.47) becomes

$$e^{-kt} \int_0^t \phi(s)ds \leq \int_0^t (a + bs)e^{-ks}ds \leq \frac{1}{k}(a + bt)(1 - e^{-kt}); \quad \forall t \in [0, T],$$

from which we obtain

$$\int_0^t \phi(s)ds \leq \frac{1}{k}(a + bt)(e^{kt} - 1); \quad \forall t \in [0, T]. \tag{5.48}$$

It then follows, via (5.48), that

$$\begin{aligned} \phi(t) &\leq a + bt + k \int_0^t \phi(s)ds \\ &\leq a + bt + (a + bt)(e^{kt} - 1) \\ &= (a + bt)e^{kt}; \quad \forall t \in [0, T], \end{aligned}$$

as required. $\square$

We now introduce the following functions $v, w : \bar{D}_T \to \mathbb{R}$.

Definition 5.7 Let $f : \mathbb{R} \to \mathbb{R}$ be such that $f \in H_\alpha$ for some $\alpha \in (0, 1]$, $u \in B_A^T$ and $\hat{u} \in B_B$. We introduce the following functions $v, w : \bar{D}_T \to \mathbb{R}$, defined by

$$v(x, t) = \int_{\lambda=-\infty}^{\infty} \hat{u}\left(x + 2\sqrt{t}\,\lambda\right)e^{-\lambda^2}d\lambda; \quad \forall (x, t) \in \bar{D}_T, \tag{5.49}$$

$$w(x, t) = \int_{\tau=0}^{t} \int_{\lambda=-\infty}^{\infty} f\left(u\left(x + 2\sqrt{t-\tau}\,\lambda, \tau\right)\right)e^{-\lambda^2}d\lambda d\tau; \quad \forall (x, t) \in \bar{D}_T. \tag{5.50}$$

Lemma† 5.8 *Let $v, w : \bar{D}_T \to \mathbb{R}$ be given as in (5.49) and (5.50). Then v and w are well defined functions and $v, w \in B_A^T$. Moreover,*

(i) $v(x, 0) = \sqrt{\pi}\hat{u}(x); \ \forall x \in \mathbb{R}$.
(ii) $w(x, 0) = 0; \ \forall x \in \mathbb{R}$.

Proof (i) $\underline{v : \bar{D}_T \to \mathbb{R}}$.
The improper integral on the right hand side of (5.49) exists for each $(x, t) \in \bar{D}_T$. Thus v is well-defined on $\bar{D}_T$. Moreover,

$$|v(x, t)| \leq \int_{-\infty}^{\infty} |\hat{u}(x + 2\sqrt{t}\,\lambda)|e^{-\lambda^2}d\lambda$$

$$\leq \sqrt{\pi}\,\|\hat{u}\|_B$$

and so v is bounded on $\bar{D}_T$. We next establish that v is continuous on $\bar{D}_T$. We define $G : [-a, a] \times [0, T] \times [\Lambda, \Lambda] \to \mathbb{R}$ by

$$G(x, t, \lambda) = \hat{u}(x + 2\sqrt{t}\,\lambda)e^{-\lambda^2}$$

for all $(x, t, \lambda) \in [-a, a] \times [0, T] \times [-\Lambda, \Lambda]$, for any fixed $a, \Lambda > 0$. G is continuous by composition. Moreover the integral

$$\int_{-\infty}^{\infty} G(x, t, \lambda)d\lambda$$

is uniformly convergent for all $(x, t) \in [-a, a] \times [0, T]$. Therefore v is continuous on $[-a, a] \times [0, T]$. This holds for any $a > 0$, and so v is continuous on $\bar{D}_T$. Thus we have shown that $v \in B_A^T$, as required. In particular, via (5.49),

$$v(x, 0) = \int_{-\infty}^{\infty} \hat{u}(x)e^{-\lambda^2}d\lambda = \sqrt{\pi}\,\hat{u}(x); \quad \forall x \in \mathbb{R}.$$

(ii) $\underline{w : \bar{D}_T \to \mathbb{R}}$.
We first observe that, with $f \in H_\alpha$ for some $\alpha \in (0, 1]$, and with $u \in B_A^T$, then

$$f(u) \in B_A^T. \tag{5.51}$$

The double integral (5.50) is then convergent for each $(x, t) \in \bar{D}_T$, and so $w : \bar{D}_T \to \mathbb{R}$ is well-defined. In particular

$$|w(x, t)| \leq \int_0^t \int_{-\infty}^{\infty} |f\left(u\left(x + 2\sqrt{t - \tau}\,\lambda, \tau\right)\right)|e^{-\lambda^2}d\lambda d\tau$$

$$\leq \|f(u)\|_A \int_0^t \int_{-\infty}^{\infty} e^{-\lambda^2}d\lambda d\tau$$

$$\leq \sqrt{\pi}\,T\|f(u)\|_A$$

and so, w is bounded on $\bar{D}_T$. We next consider

$$h : \{(x,t,\tau) \in \mathbb{R}^3 : (x,t) \in [-a,a] \times [0,T], \tau \in [0,t]\} \to \mathbb{R}$$

for any $a > 0$, with

$$h(x,t,\tau) = \int_{-\infty}^{\infty} f\left(u\left(x + 2\sqrt{t-\tau}\,\lambda, \tau\right)\right) e^{-\lambda^2} d\lambda. \qquad (5.52)$$

Now, $f\left(u\left(x + 2\sqrt{t-\tau}\lambda, \tau\right)\right)$ is continuous for all

$$(x,t,\tau,\lambda) \in \{(x,t,\tau) \in \mathbb{R}^3 : (x,t) \in [-a,a] \times [0,T], \tau \in [0,t]\} \times [-\Lambda, \Lambda]$$

(for any $\Lambda > 0$), and the integral in (5.52) is uniformly convergent for all

$$(x,t,\tau) \in \{(x,t,\tau) \in \mathbb{R}^3 : (x,t) \in [-a,a] \times [0,T], \tau \in [0,t]\}.$$

Thus $h(x,t,\tau)$ is a continuous function on

$$\{(x,t,\tau) \in \mathbb{R}^3 : (x,t) \in [-a,a] \times [0,T], \tau \in [0,t]\}.$$

Now observe

$$w(x,t) = \int_0^t h(x,t,\tau) d\tau \qquad (5.53)$$

for all $(x,t) \in \bar{D}_T$. Since $h(x,t,\tau)$ is continuous for all

$$(x,t,\tau) \in \{(x,t,\tau) \in \mathbb{R}^3 : (x,t) \in [-a,a] \times [0,T], \tau \in [0,t]\},$$

then it follows that $w(x,t)$ is continuous for all $(x,t) \in [-a,a] \times [0,T]$. Thus w is continuous on $\bar{D}_T$ and we conclude that $w \in B_A^T$.

Finally,

$$w(x,0) = \int_0^0 h(x,0,0) d\tau = 0; \quad \forall x \in \mathbb{R}.$$

The proof is complete. $\qquad\qquad\qquad\qquad\qquad\qquad\qquad\qquad\qquad\qquad \square$

With $(x,t) \in D_T$, the expression (5.49) for v may be re-written via simple substitution. For $(x,t) \in D_T$, we make the substitution $s = x + 2\sqrt{t}\lambda$ in (5.49), after which we obtain

$$v(x,t) = \frac{1}{2\sqrt{t}} \int_{-\infty}^{\infty} \hat{u}(s) e^{-\frac{(x-s)^2}{4t}} ds; \quad \forall (x,t) \in D_T. \qquad (5.54)$$

We next introduce $F : \bar{D}_T \to \mathbb{R}$ such that

$$F(x,t) = f(u(x,t)); \quad \forall (x,t) \in \bar{D}_T. \qquad (5.55)$$

Now, since $f \in H_\alpha$ and $u \in B_A^T$, then F is bounded and continuous on $\bar{D}_T$. It then follows from Section 5.1, that we may write

$$w(x,t) = \lim_{n\to\infty} \int_{\tau=0}^{t-1/n} \int_{-\infty}^{\infty} \frac{f(u(s,\tau))}{2\sqrt{t-\tau}} e^{-\frac{(x-s)^2}{4(t-\tau)}} ds \, d\tau; \quad \forall (x,t) \in D_T. \tag{5.56}$$

We can now state:

Lemma† 5.9 (Regularity) *The functions* $v, w \in B_A^T$ *are such that* v_t, v_x, v_{xx} *and* w_x, *all exist and are continuous on* D_T. *Moreover, the derivatives are given by*

$$v_x(x,t) = \frac{1}{t^{\frac{1}{2}}} \int_{-\infty}^{\infty} \hat{u}\left(x + 2\sqrt{t}w\right) w e^{-w^2} dw,$$

$$v_t(x,t) = v_{xx}(x,t) = \frac{1}{t} \int_{-\infty}^{\infty} \hat{u}\left(x + 2\sqrt{t}w\right)(w^2 - 1/2) e^{-w^2} dw,$$

$$w_x(x,t) = \lim_{n\to\infty} \int_0^{t-1/n} \int_{-\infty}^{\infty} \frac{f\left(u\left(x + 2\sqrt{t-\tau}w, \tau\right)\right)}{(t-\tau)^{\frac{1}{2}}} w e^{-w^2} dw d\tau; \quad \forall (x,t) \in D_T.$$

Suppose also that u_x *exists and is bounded on* D_T, *then* w_t *and* w_{xx} *also exist and are continuous on* D_T, *with*

$$w_{xx}(x,t) = \lim_{n\to\infty} \int_0^{t-1/n} \int_{-\infty}^{\infty} \frac{f\left(u\left(x + 2\sqrt{t-\tau}w, \tau\right)\right)}{(t-\tau)}(w^2 - 1/2) e^{-w^2} dw d\tau,$$

$$w_t(x,t) = \lim_{n\to\infty} \int_0^{t-1/n} \int_{-\infty}^{\infty} \frac{f\left(u\left(x + 2\sqrt{t-\tau}w, \tau\right)\right)}{(t-\tau)}(w^2 - 1/2) e^{-w^2} dw d\tau$$
$$+ \sqrt{\pi} f(u(x,t)); \quad \forall (x,t) \in D_T. \tag{5.57}$$

Proof We first give a proof for v. We introduce the function $\phi : [-a, a] \times [t_0, T] \times (-\infty, \infty) \to \mathbb{R}$, given by

$$\phi(x,t,s) = \frac{\hat{u}(s)}{2\sqrt{t}} e^{-\frac{(x-s)^2}{4t}} \tag{5.58}$$

for $(x,t,s) \in [-a, a] \times [t_0, T] \times (-\infty, \infty)$ (for any $a > 0$ and $0 < t_0 < T$). Then,

$$v(x,t) = \int_{-\infty}^{\infty} \phi(x,t,s) ds$$

on $[-a, a] \times [t_0, T]$. Now, an examination of (5.58) shows that ϕ_t, ϕ_x and ϕ_{xx} all exist and are continuous on $[-a, a] \times [t_0, T] \times (-\infty, \infty)$, whilst the improper integrals

$$\int_{-\infty}^{\infty} \phi_x(x,t,s)ds, \quad \int_{-\infty}^{\infty} \phi_t(x,t,s)ds, \quad \int_{-\infty}^{\infty} \phi_{xx}(x,t,s)ds$$

are uniformly convergent for all $(x,t) \in [-a,a] \times [t_0, T]$. It follows that v_t, v_x and v_{xx} all exist and are continuous on $[-a,a] \times [t_0, T]$, for any $a > 0$ and $0 < t_0 < T$. Thus v_t, v_x and v_{xx} all exist and are continuous on D_T. Moreover,

$$v_x = \int_{-\infty}^{\infty} \phi_x(x,t,s)ds, \quad v_t = \int_{-\infty}^{\infty} \phi_t(x,t,s)ds, \quad v_{xx} = \int_{-\infty}^{\infty} \phi_{xx}(x,t,s)ds.$$
(5.59)

The given derivatives are now obtained by replacing ϕ_t, ϕ_x and ϕ_{xx} in the above, followed by the substitution $s = x + 2\sqrt{t}w$.

We now give a proof for w. First we recall that $f \in H_\alpha$ and $u \in B_A^T$ so that $f(u)$ is bounded and continuous on $\bar{D}_T$. It then follows, via Theorem 5.1, that w_x exists and is continuous on D_T, and the derivative formula follows via (5.8). Next, when $u \in B_A^T$ is such that u_x exists and is bounded on D_T, it follows with $f \in H_\alpha$, that $f(u)$ satisfies condition (H) (in Section 5.1) on $\bar{D}_T$ (via an application of the mean value theorem). It then follows from Theorem 5.2 that w_t and w_{xx} exist and are continuous on D_T. The derivative formulae follow from (5.9), (5.10) and (5.24).   □

5.3 Equivalence Lemma and Integral Equation

We relate solutions of an associated *integral equation* to solutions of (B-R-D-C). We have,

Lemma 5.10 (Hölder Equivalence) *Let* $f \in H_\alpha$ *for some* $\alpha \in (0, 1]$ *and* $u_0 \in BPC^2(\mathbb{R})$. *Then, the following statements are equivalent:*

(a) $u \in B_A^T$ *and* $u : \bar{D}_T \to \mathbb{R}$ *satisfies the integral equation*

$$u(x,t) = \frac{1}{\sqrt{\pi}} \int_{-\infty}^{\infty} u_0\left(x + 2\sqrt{t}\lambda\right) e^{-\lambda^2} d\lambda$$

$$+ \frac{1}{\sqrt{\pi}} \int_0^t \int_{-\infty}^{\infty} f\left(u\left(x + 2\sqrt{t-\tau}\lambda, \tau\right)\right) e^{-\lambda^2} d\lambda d\tau$$

for all $(x,t) \in \bar{D}_T$.
(b) $u : \bar{D}_T \to \mathbb{R}$ *is a solution to* (B-R-D-C) *on* $\bar{D}_T$.

Proof (i) <u>(a)⇒(b)</u>.
Suppose (a) holds for $u : \bar{D}_T \to \mathbb{R}$ with $u \in B_A^T$ (note that the right hand side of the integral equation in (a) is well-defined as a function in B_A^T for any

$u \in B_A^T$, via Lemma 5.8, since $u_0 \in B_B$). In particular, via Lemma 5.8, we have

$$u(x, 0) = u_0(x); \ \forall x \in \mathbb{R}, \tag{5.60}$$

whilst

$$u(x, t) \text{ is uniformly bounded as } |x| \to \infty \text{ for } t \in [0, T] \tag{5.61}$$

as $u \in B_A^T$. Now we have, via Lemma 5.9, (4.9), (4.10), Theorem 5.1 and (a) that u_x exists and is bounded on D_T. It then follows, again via Lemma 5.9, that u_t, u_x and u_{xx} all exist and are continuous on D_T. Finally using the derivative formula given in Lemma 5.9, a direct substitution shows that

$$u_t - u_{xx} - f(u) = 0 \text{ on } D_T. \tag{5.62}$$

Together, (5.60), (5.61) and (5.62) imply that $u : \bar{D}_T \to \mathbb{R}$ is a solution to (B-R-D-C) on $\bar{D}_T$.

(ii) (b)$\Rightarrow$(a).
Let $u : \bar{D}_T \to \mathbb{R}$ be a solution of (B-R-D-C) on $\bar{D}_T$. Then $f \circ u : \bar{D}_T \to \mathbb{R}$ is bounded and continuous. The result then follows immediately from Theorem 4.9. □

Lemma† 5.11 (Integral Equation for $f \in L_u$) *Let $f \in L_u$ and $u_0 \in BPC^2(\mathbb{R})$, and let $u : \bar{D}_T \to \mathbb{R}$ be a solution to (B-R-D-C) on $\bar{D}_T$. Then,*

$$u(x, t) = \frac{1}{\sqrt{\pi}} \int_{-\infty}^{\infty} u_0 \left(x + 2\sqrt{t}\lambda \right) e^{-\lambda^2} d\lambda$$
$$+ \frac{1}{\sqrt{\pi}} \int_0^t \int_{-\infty}^{\infty} f \left(u \left(x + 2\sqrt{t - \tau}\lambda, \tau \right) \right) e^{-\lambda^2} d\lambda d\tau$$

for all $(x, t) \in \bar{D}_T$.

Proof Let $u : \bar{D}_T \to \mathbb{R}$ be a solution to (B-R-D-C) on $\bar{D}_T$. Then, since $f \in L_u$, we conclude that $f \circ u : \bar{D}_T \to \mathbb{R}$ is bounded and continuous. The result then follows immediately from Theorem 4.9. □

5.4 Derivative Estimates

We now move on to establishing derivative bounds for solutions to (B-R-D-C) on $\bar{D}_T$ throughout we take $u_0 \in BPC^2(\mathbb{R})$. We first need the following:

Lemma‡ 5.12 (Derivative Estimate) *Let $f \in H_\alpha$ for some $\alpha \in (0, 1]$ and $u : \bar{D}_T \to \mathbb{R}$ be a solution to (B-R-D-C) on $\bar{D}_T$. Then,*

$$|u_x(x, t)| \le \frac{2M_T}{\sqrt{\pi}}(1 + T^{\frac{1}{2}}) + M_0'; \quad \forall (x, t) \in D_T,$$

where $M_0' > 0$ is an upper bound for $|u_0'| : \mathbb{R} \to \mathbb{R}$ and $M_T > 0$ is an upper bound for $|f \circ u| : \bar{D}_T \to \mathbb{R}$.

Proof Let $u : \bar{D}_T \to \mathbb{R}$ be a solution to (B-R-D-C) on $\bar{D}_T$. Then, via Lemma 5.10 and Lemma 5.9, for any $(x, t) \in D_T$,

$$\begin{aligned}
u_x(x, t) &= \left(\frac{1}{\sqrt{\pi}} \int_{-\infty}^{\infty} u_0 \left(x + 2\sqrt{t}\lambda \right) e^{-\lambda^2} d\lambda \right)_x \\
&\quad + \left(\frac{1}{\sqrt{\pi}} \int_0^t \int_{-\infty}^{\infty} f \left(u \left(x + 2\sqrt{t - \tau}\lambda, \tau \right) \right) e^{-\lambda^2} d\lambda d\tau \right)_x \\
&= \frac{1}{\sqrt{\pi} t^{\frac{1}{2}}} \int_{-\infty}^{\infty} u_0 \left(x + 2\sqrt{t}\lambda \right) \lambda e^{-\lambda^2} d\lambda \\
&\quad + \lim_{n\to\infty} \frac{1}{\sqrt{\pi}} \int_0^{t-1/n} \int_{-\infty}^{\infty} \frac{f \left(u \left(x + 2\sqrt{t - \tau}\lambda, \tau \right) \right)}{(t - \tau)^{\frac{1}{2}}} \lambda e^{-\lambda^2} d\lambda d\tau \\
&= \frac{1}{\sqrt{\pi}} \int_{-\infty}^{\infty} u_0' \left(x + 2\sqrt{t}\lambda \right) e^{-\lambda^2} d\lambda \\
&\quad + \lim_{n\to\infty} \frac{1}{\sqrt{\pi}} \int_0^{t-1/n} \int_{-\infty}^{\infty} \frac{f \left(u \left(x + 2\sqrt{t - \tau}\lambda, \tau \right) \right)}{(t - \tau)^{\frac{1}{2}}} \lambda e^{-\lambda^2} d\lambda d\tau
\end{aligned}$$

following an integration by parts. It follows that, for any $(x, t) \in D_T$,

$$\begin{aligned}
|u_x&(x, t)| \\
&\le \frac{1}{\sqrt{\pi}} \int_{-\infty}^{\infty} |u_0' \left(x + 2\sqrt{t}\lambda \right)| e^{-\lambda^2} d\lambda \\
&\quad + \lim_{n\to\infty} \frac{1}{\sqrt{\pi}} \left| \int_0^{t-1/n} \int_{-\infty}^{\infty} \frac{f \left(u \left(x + 2\sqrt{t - \tau}\lambda, \tau \right) \right)}{(t - \tau)^{\frac{1}{2}}} \lambda e^{-\lambda^2} d\lambda d\tau \right| \\
&\le M_0' + \lim_{n\to\infty} \frac{1}{\sqrt{\pi}} \left| \int_0^{t-1/n} \int_{-\infty}^{\infty} \frac{f \left(u \left(x + 2\sqrt{t - \tau}\lambda, \tau \right) \right)}{(t - \tau)^{\frac{1}{2}}} \lambda e^{-\lambda^2} d\lambda d\tau \right|.
\end{aligned}$$

$$(5.63)$$

Now, for any $(x, t) \in D_T$ and $0 < 1/n < min\{1, t\}$,

$$\left| \int_0^{t-1/n} \int_{-\infty}^{\infty} \frac{f\left(u\left(x + 2\sqrt{t-\tau}\lambda, \tau\right)\right)}{(t-\tau)^{\frac{1}{2}}} \lambda e^{-\lambda^2} d\lambda d\tau \right|$$

$$\leq \int_0^{t-1/n} \int_{-\infty}^{\infty} \frac{\left|f\left(u\left(x + 2\sqrt{t-\tau}\lambda, \tau\right)\right)\right|}{(t-\tau)^{\frac{1}{2}}} |\lambda| e^{-\lambda^2} d\lambda d\tau$$

$$\leq M_T \left(\int_0^{t-1/n} \frac{1}{(t-\tau)^{\frac{1}{2}}} d\tau \right) \left(\int_{-\infty}^{\infty} |\lambda| e^{-\lambda^2} d\lambda \right)$$

$$= 2 M_T (t^{\frac{1}{2}} - (1/n)^{\frac{1}{2}})$$

$$\leq 2 M_T (T^{\frac{1}{2}} + 1). \tag{5.64}$$

It follows from (5.63) and (5.64), that

$$|u_x(x, t)| \leq M_0' + \frac{2M_T}{\sqrt{\pi}} (1 + T^{\frac{1}{2}}); \quad \forall (x, t) \in D_T,$$

as required. $\qquad \qquad \square$

Remark 5.13 Observe that the proof of Lemma 5.12 only requires that the solution $u : \bar{D}_T \to \mathbb{R}$ satisfies an integral equation as in Lemma 5.10 or Lemma 5.11. Therefore Lemma 5.12 can also be established for $f \in L_u$. However, since subsequent applications of this derivative estimate only concern $f \in H_\alpha$, it is stated as above. $\qquad \lrcorner$

We next have,

Lemma‡ 5.14 *Let* $f \in H_\alpha$ *for some* $\alpha \in (0, 1]$ *and let* $u : \bar{D}_T \to \mathbb{R}$ *be a solution to (B-R-D-C) on* $\bar{D}_T$. *Then* $f \circ u : \bar{D}_T \to \mathbb{R}$ *satisfies*

$$|f(u(y, t)) - f(u(x, t))| \leq k_T |y - x|^\alpha; \quad \forall (x, t), (y, t) \in \bar{D}_T$$

where

$$k_T = k_E \left(\frac{2M_T}{\sqrt{\pi}} (1 + T^{\frac{1}{2}}) + M_0' \right)^\alpha$$

and $k_E > 0$ *is a Hölder constant for* $f : \mathbb{R} \to \mathbb{R}$ *on the closed bounded interval* $[-U_T, U_T]$, *with* $U_T > 0$ *being an upper bound for* $|u| : \bar{D}_T \to \mathbb{R}$.

Proof Let $(x, t), (y, t) \in D_T$, then $u(x, t), u(y, t) \in [-U_T, U_T]$, and so, since $f \in H_\alpha$, then

$$|f(u(y, t)) - f(u(x, t))| \leq k_E |u(y, t) - u(x, t)|^\alpha \tag{5.65}$$

where $k_E > 0$ is a Hölder constant for $f : \mathbb{R} \to \mathbb{R}$ on the closed bounded interval $[-U_T, U_T]$. However, it follows from the mean value theorem together with Lemma 5.12, that

$$|u(y, t) - u(x, t)| \leq \left(\frac{2M_T}{\sqrt{\pi}}(1 + T^{\frac{1}{2}}) + M_0' \right) |y - x|. \qquad (5.66)$$

Combining (5.65) and (5.66) we obtain

$$|f(u(y, t)) - f(u(x, t))| \leq k_T |y - x|^\alpha \quad \forall\, (x, t), (y, t) \in D_T, \qquad (5.67)$$

with

$$k_T = k_E \left(\frac{2M_T}{\sqrt{\pi}}(1 + T^{\frac{1}{2}}) + M_0' \right)^\alpha.$$

Now, for fixed $x, y \in \mathbb{R}$, the left-hand side of (5.67) is continuous for $t \in [0, T]$, whilst the right-hand side of (5.67) is independent of t. It follows that the inequality (5.67) extends from D_T onto $\bar{D}_T$, and the proof is complete. $\qquad \square$

We are now in a position to state,

Lemma$_\ddagger$ 5.15 (Derivative Estimates) *Let $f \in H_\alpha$ for some $\alpha \in (0, 1]$ and $u : \bar{D}_T \to \mathbb{R}$ be a solution to (B-R-D-C) on $\bar{D}_T$. Then,*

$$|u_{xx}(x, t)| \leq \frac{2^{\alpha+1} I_\alpha}{\alpha\sqrt{\pi}} k_T (1 + T^{\frac{1}{2}\alpha}) + M_0''; \quad \forall (x, t) \in D_T,$$

$$|u_t(x, t)| \leq \frac{2^{\alpha+1} I_\alpha}{\alpha\sqrt{\pi}} k_T (1 + T^{\frac{1}{2}\alpha}) + M_0'' + M_T; \quad \forall (x, t) \in D_T,$$

where $M_0'' > 0$ is an upper bound for $|u_0''| : \mathbb{R} \to \mathbb{R}$ and

$$I_\alpha = \int_{-\infty}^{\infty} |\lambda|^\alpha |\lambda^2 - 1/2| e^{-\lambda^2} d\lambda \quad (> 0).$$

Proof Let $u : \bar{D}_T \to \mathbb{R}$ be a solution to (B-R-D-C) on $\bar{D}_T$. Then u_x exists and is bounded on D_T, via Lemma 5.12. It then follows, via Lemma 5.9 and Lemma 5.10, for any $(x, t) \in D_T$,

$u_{xx}(x, t)$

$$= \left(\frac{1}{\sqrt{\pi}} \int_{-\infty}^{\infty} u_0 \left(x + 2\sqrt{t}\lambda \right) e^{-\lambda^2} d\lambda \right)_{xx}$$

$$+ \left(\frac{1}{\sqrt{\pi}} \int_0^t \int_{-\infty}^{\infty} f \left(u \left(x + 2\sqrt{t - \tau}\lambda, \tau \right) \right) e^{-\lambda^2} d\lambda d\tau \right)_{xx}$$

$$= \frac{1}{\sqrt{\pi t}} \int_{-\infty}^{\infty} u_0 \left(x + 2\sqrt{t}\lambda \right) (\lambda^2 - 1/2) e^{-\lambda^2} d\lambda$$

$$+ \lim_{n \to \infty} \frac{1}{\sqrt{\pi}} \int_0^{t-1/n} \int_{-\infty}^{\infty} \frac{f \left(u \left(x + 2\sqrt{t-\tau}\lambda, \tau \right) \right)}{(t-\tau)} (\lambda^2 - 1/2) e^{-\lambda^2} d\lambda d\tau$$

$$= \frac{1}{\sqrt{\pi}} \int_{-\infty}^{\infty} u_0'' \left(x + 2\sqrt{t}\lambda \right) e^{-\lambda^2} d\lambda$$

$$+ \lim_{n \to \infty} \frac{1}{\sqrt{\pi}} \int_0^{t-1/n} \int_{-\infty}^{\infty} \frac{f \left(u \left(x + 2\sqrt{t-\tau}\lambda, \tau \right) \right)}{(t-\tau)} (\lambda^2 - 1/2) e^{-\lambda^2} d\lambda d\tau$$

following an integration by parts. Thus, for any $(x, t) \in D_T$,

$$|u_{xx}(x, t)| \le \frac{1}{\sqrt{\pi}} \int_{-\infty}^{\infty} |u_0'' \left(x + 2\sqrt{t}\lambda \right)| e^{-\lambda^2} d\lambda$$

$$+ \lim_{n \to \infty} \frac{1}{\sqrt{\pi}} \left| \int_0^{t-1/n} \int_{-\infty}^{\infty} \frac{f \left(u \left(x + 2\sqrt{t-\tau}\lambda, \tau \right) \right)}{(t-\tau)} \right.$$

$$\left. (\lambda^2 - 1/2) e^{-\lambda^2} d\lambda d\tau \right|$$

$$\le M_0'' + \lim_{n \to \infty} \frac{1}{\sqrt{\pi}} \left| \int_0^{t-1/n} \int_{-\infty}^{\infty} \frac{f \left(u \left(x + 2\sqrt{t-\tau}\lambda, \tau \right) \right)}{(t-\tau)} \right.$$

$$\left. (\lambda^2 - 1/2) e^{-\lambda^2} d\lambda d\tau \right|. \qquad (5.68)$$

Now, for any $(x, t) \in D_T$ and $0 < 1/n < min\{1, t\}$,

$$\frac{1}{\sqrt{\pi}} \left| \int_0^{t-1/n} \int_{-\infty}^{\infty} \frac{f \left(u \left(x + 2\sqrt{t-\tau}\lambda, \tau \right) \right)}{(t-\tau)} (\lambda^2 - 1/2) e^{-\lambda^2} d\lambda d\tau \right|$$

$$\le \frac{1}{\sqrt{\pi}} \left| \int_0^{t-1/n} \int_{-\infty}^{\infty} \frac{f \left(u \left(x + 2\sqrt{t-\tau}\lambda, \tau \right) \right) - f(u(x, \tau))}{(t-\tau)} \right.$$

$$\left. (\lambda^2 - 1/2) e^{-\lambda^2} d\lambda d\tau \right|$$

$$+ \frac{1}{\sqrt{\pi}} \left| \int_0^{t-1/n} \int_{-\infty}^{\infty} \frac{f(u(x, \tau))(\lambda^2 - 1/2)}{(t-\tau)} e^{-\lambda^2} d\lambda d\tau \right| \qquad (5.69)$$

via the triangle inequality. However, the second term on the right-hand side of (5.69) vanishes, and so

$$\frac{1}{\sqrt{\pi}} \left| \int_0^{t-1/n} \int_{-\infty}^{\infty} \frac{f\left(u\left(x + 2\sqrt{t-\tau}\lambda, \tau\right)\right)}{(t-\tau)}(\lambda^2 - 1/2)e^{-\lambda^2}d\lambda d\tau \right|$$

$$\leq \frac{1}{\sqrt{\pi}} \left| \int_0^{t-1/n} \int_{-\infty}^{\infty} \frac{f\left(u\left(x + 2\sqrt{t-\tau}\lambda, \tau\right)\right) - f(u(x,\tau))}{(t-\tau)} \right.$$

$$\left. (\lambda^2 - 1/2)e^{-\lambda^2}d\lambda d\tau \right|$$

$$\leq \frac{1}{\sqrt{\pi}} \int_0^{t-1/n} \int_{-\infty}^{\infty} \frac{|f\left(u\left(x + 2\sqrt{t-\tau}\lambda, \tau\right)\right) - f(u(x,\tau))|}{(t-\tau)}$$

$$|\lambda^2 - 1/2|e^{-\lambda^2}d\lambda d\tau$$

$$\leq \frac{2^\alpha k_T}{\sqrt{\pi}} \int_0^{t-1/n} \int_{-\infty}^{\infty} \frac{|\lambda|^\alpha |\lambda^2 - 1/2|}{(t-\tau)^{1-\alpha/2}} e^{-\lambda^2}d\lambda d\tau \quad \text{(via Lemma 5.14)}$$

$$(5.70)$$

$$\leq \frac{2^\alpha I_\alpha k_T}{\sqrt{\pi}} \int_0^{t-1/n} \frac{1}{(t-\tau)^{1-\alpha/2}}d\tau$$

$$\leq \frac{2^{\alpha+1} I_\alpha k_T}{\alpha\sqrt{\pi}} (t^{\alpha/2} - 1/n^{\alpha/2})$$

$$\leq \frac{2^{\alpha+1} I_\alpha k_T}{\alpha\sqrt{\pi}} (1 + T^{\alpha/2}), \tag{5.71}$$

where

$$I_\alpha = \int_{-\infty}^{\infty} |\lambda|^\alpha |\lambda^2 - 1/2|e^{-\lambda^2}d\lambda \quad (> 0). \tag{5.72}$$

It follows from (5.68) and (5.71) that

$$|u_{xx}(x,t)| \leq M_0'' + \frac{2^{\alpha+1} I_\alpha k_T}{\alpha\sqrt{\pi}} (1 + T^{\frac{\alpha}{2}}); \quad \forall(x,t) \in D_T, \tag{5.73}$$

as required. Now, since $u : \bar{D}_T \to \mathbb{R}$ is a solution to (B-R-D-C) on $\bar{D}_T$, then

$$u_t(x,t) = u_{xx}(x,t) + f(u(x,t)); \quad \forall(x,t) \in D_T, \tag{5.74}$$

via Definition 2.1. Thus, via the triangle inequality and (5.73),

$$|u_t(x,t)| \leq |u_{xx}(x,t)| + |f(u(x,t))|$$

$$\leq M_T + M_0'' + \frac{2^{\alpha+1} I_\alpha k_T (1 + T^{\frac{\alpha}{2}})}{\alpha\sqrt{\pi}}; \quad \forall(x,t) \in D_T, \tag{5.75}$$

as required. □

An additional useful result is,

Corollary‡ 5.16 *Let $f \in H_\alpha$ for some $\alpha \in (0, 1]$ and $u : \bar{D}_T \to \mathbb{R}$ be a solution to (B-R-D-C) on $\bar{D}_T$. Then u is uniformly continuous on $\bar{D}_T$.*

Proof It follows from Lemma 5.12 and Lemma 5.15 together with the mean value theorem, that

$$|u(x_2, t_2) - u(x_1, t_1)| \leq M(|x_2 - x_1| + |t_2 - t_1|), \quad \forall (x_2, t_2), (x_1, t_1) \in D_T,$$

with $M > 0$ being the maximum of the derivative bounds for u_x and u_t on D_T. Since u is continuous on $\bar{D}_T$, it follows that

$$|u(x_2, t_2) - u(x_1, t_1)| \leq M(|x_2 - x_1| + |t_2 - t_1|), \quad \forall (x_2, t_2), (x_1, t_1) \in \bar{D}_T,$$

and the result follows.   □

Remark 5.17 For fixed $\alpha \in (0, 1]$ and $T > 0$, Lemma 5.12 and Lemma 5.15 establish that for any solution $u : \bar{D}_T \to \mathbb{R}$ of (B-R-D-C) on $\bar{D}_T$, then $|u_t|$, $|u_x|$ and $|u_{xx}|$ are bounded on D_T, with bounds which only depend upon α, T, $\|u_0'\|_B$, $\|u_0''\|_B$, $\|f \circ u\|_A$, $\|u\|_A$ and a Hölder constant for f on $[-U_T, U_T]$ with U_T being an upper bound for $\|u\|_A$. Such results are often referred to as *Schauder Estimates* for (B-R-D-C) (after J. Schauder [68], [69] of whose results for elliptic problems were extended to parabolic problems by A. Friedman, [21]). For additional information concerning the development of these types of results, see [26] and [62].   ⌐

6

The Bounded Reaction-Diffusion Cauchy Problem with $f \in L$

This chapter contains the classical global well-posedness result for a priori bounded (B-R-D-C) with $f \in L$ and $u_0 \in \text{BPC}^2(\mathbb{R})$. Uniqueness is established via an application of a maximum principle, such as Theorem 3.6, which is also used to prove a Local Existence and Uniqueness result, together with an application of the Banach fixed point theorem. This result is extended to a Global Existence and Uniqueness result for a priori bounded (B-R-D-C). A continuous dependence result is obtained via an application of the generalised Gronwall's inequality (Proposition 5.6) to the integral equation in the Hölder Equivalence Lemma 5.10. These results are combined to obtain the global well-posedness result. It should be noted that all results marked with subscript † in this chapter are adapted from standard results in [56]. Throughout, we take $u_0 \in \text{BPC}^2(\mathbb{R})$. To begin,

Theorem† **6.1** (Uniqueness) *Let* $f \in L$. *Then (B-R-D-C) has at most one solution on* $\bar{D}_T$ *for any* $T > 0$.

Proof Let $u^{(1)} : \bar{D}_T \to \mathbb{R}$ and $u^{(2)} : \bar{D}_T \to \mathbb{R}$ both be solutions to (B-R-D-C) on $\bar{D}_T$. We now define the function $h : \bar{D}_T \to \mathbb{R}$ by

$$h(x,t) = \begin{cases} \frac{f(u^{(2)}(x,t)) - f(u^{(1)}(x,t))}{(u^{(2)}(x,t) - u^{(1)}(x,t))} & ; u^{(2)}(x,t) \neq u^{(1)}(x,t) \\ 0 & ; u^{(2)}(x,t) = u^{(1)}(x,t). \end{cases} \tag{6.1}$$

Since $u^{(2)}$ and $u^{(1)}$ are bounded on $\bar{D}_T$, it follows since $f \in L$, that h is bounded on $\bar{D}_T$. Next introduce $w : \bar{D}_T \to \mathbb{R}$ such that

$$w(x,t) = u^{(2)}(x,t) - u^{(1)}(x,t); \quad \forall (x,t) \in \bar{D}_T. \tag{6.2}$$

Thus w is bounded and continuous on $\bar{D}_T$, and such that w_t, w_x and w_{xx} all exist and are continuous on D_T. Moreover,

$$w_t - w_{xx} - h(x,t)w = 0 \text{ on } D_T,$$

$$w = 0 \text{ on } \partial D. \tag{6.3}$$

It follows from (6.3) and Theorem 3.6 that $w \leq 0$ on $\bar{D}_T$, and so, $u^{(1)} \geq u^{(2)}$ on $\bar{D}_T$. On considering $w' : \bar{D}_T \to \mathbb{R}$ defined as

$$w'(x, t) = u^{(1)}(x, t) - u^{(2)}(x, t); \quad \forall (x, t) \in \bar{D}_T,$$

it follows by a symmetrical argument that $u^{(2)} \geq u^{(1)}$. Hence $u^{(1)} = u^{(2)}$, as required. □

We can now state the first existence and uniqueness result as follows,

Theorem† 6.2 (Local Existence and Uniqueness) *Let $f \in L$. Then (B-R-D-C) has a unique solution on $\bar{D}_\delta$, where*

$$\delta = \frac{1}{2k_0 + |f(0)|}$$

and $k_0 > 0$ is a Lipschitz constant for $f : \mathbb{R} \to \mathbb{R}$ on the interval $[-\sigma, \sigma]$. Here $\sigma = 2\|u_0\|_B + 1$. Moreover $\|u\|_A \leq 2\|u_0\|_B + 1$ on $\bar{D}_\delta$.

Proof We consider the closed, bounded subset $\hat{B}_A^\delta$ of the Banach space B_A^δ, equipped with $\| \cdot \|_A$, where

$$\hat{B}_A^\delta = \{v : \bar{D}_\delta \to \mathbb{R} : v \in B_A^\delta \text{ and } \|v\|_A \leq 2\|u_0\|_B + 1\}. \tag{6.4}$$

We note that $\hat{B}_A^\delta$ is complete as it is a closed, bounded subset of B_A^δ. Here,

$$\delta = \frac{1}{2k_0 + |f(0)|} \leq \frac{1}{2k_0},$$

with k_0 being a Lipschitz constant for $f : \mathbb{R} \to \mathbb{R}$ on the interval $[-\sigma, \sigma]$, where $\sigma = 2\|u_0\|_B + 1$. That is,

$$|f(X) - f(Y)| \leq k_0|X - Y|; \quad \forall X, Y \in [-\sigma, \sigma]. \tag{6.5}$$

We now define the following mapping $G : \hat{B}_A^\delta \to B_A^\delta$ such that

$$G(v) = \frac{1}{\sqrt{\pi}} \int_{-\infty}^{\infty} u_0 \left(x + 2\sqrt{t}\lambda\right) e^{-\lambda^2} d\lambda$$
$$+ \frac{1}{\sqrt{\pi}} \int_0^t \int_{-\infty}^{\infty} f \left(v \left(x + 2\sqrt{t - \tau}\lambda, \tau\right)\right) e^{-\lambda^2} d\lambda d\tau \tag{6.6}$$

for any $v \in \hat{B}_A^\delta$. Via Lemma 5.8 G is well-defined as a mapping from $\hat{B}_A^\delta$ into B_A^δ. We must now show that $Im(G) \subseteq \hat{B}_A^\delta$. Let $v \in \hat{B}_A^\delta$, then

$$\|G(v)\|_A \leq \frac{1}{\sqrt{\pi}} \left\| \int_{-\infty}^{\infty} u_0 \left(x + 2\sqrt{t}\lambda\right) e^{-\lambda^2} d\lambda \right\|_A$$
$$+ \frac{1}{\sqrt{\pi}} \left\| \int_0^t \int_{-\infty}^{\infty} f \left(v \left(x + 2\sqrt{t - \tau}\lambda, \tau\right)\right) e^{-\lambda^2} d\lambda d\tau \right\|_A$$

$$\leq \frac{1}{\sqrt{\pi}} \sup_{(x,t) \in \bar{D}_\delta} \left\{ \int_{-\infty}^{\infty} |u_0(x + 2\sqrt{t}\lambda)| e^{-\lambda^2} d\lambda \right\}$$

$$+ \frac{1}{\sqrt{\pi}} \sup_{(x,t) \in \bar{D}_\delta} \left\{ \int_0^t \int_{-\infty}^{\infty} |f(v(x + 2\sqrt{t - \tau}\lambda, \tau))| e^{-\lambda^2} d\lambda d\tau \right\}$$

$$\leq \|u_0\|_B + \delta \|f(v)\|_A. \tag{6.7}$$

However, $v \in \hat{B}_A^\delta$ so that $v(x, t) \in [-\sigma, \sigma]$ for all $(x, t) \in \bar{D}_\delta$. Hence,

$$|f(v(x,t))| \leq |f(v(x,t)) - f(0)| + |f(0)| \leq k_0 |v(x,t)| + |f(0)|; \quad \forall (x, t) \in \bar{D}_\delta$$

via (6.5). Hence (recalling $f(v) \in B_A^\delta$)

$$\|f(v)\|_A \leq k_0 \|v\|_A + |f(0)|.$$

Thus (6.7) leads to

$$\|G(v)\|_A \leq \|u_0\|_B + \delta(k_0 \|v\|_A + |f(0)|)$$

$$\leq \|u_0\|_B + \delta(k_0(2\|u_0\|_B + 1) + |f(0)|) \text{ as } v \in \hat{B}_A^\delta$$

$$= \|u_0\|_B + \frac{k_0(2\|u_0\|_B + 1) + |f(0)|}{2k_0 + |f(0)|}$$

$$= \|u_0\|_B + \frac{2k_0\|u_0\|_B}{2k_0 + |f(0)|} + \frac{k_0 + |f(0)|}{2k_0 + |f(0)|}$$

$$\leq \|u_0\|_B + \|u_0\|_B + 1$$

$$= 2\|u_0\|_B + 1.$$

Therefore, by definition

$$G(v) \in \hat{B}_A^\delta; \quad \forall v \in \hat{B}_A^\delta,$$

and so $Im(G) \subseteq \hat{B}_A^\delta$, as required. Next we show that $G : \hat{B}_A^\delta \to \hat{B}_A^\delta$ is a *Contraction Mapping*. For any $v, w \in \hat{B}_A^\delta$, we have

$$\|G(v) - G(w)\|_A$$

$$= \frac{1}{\sqrt{\pi}} \left\| \int_0^t \int_{-\infty}^{\infty} [f(v(x + 2\sqrt{t - \tau}\lambda, \tau)) - f(w(x + 2\sqrt{t - \tau}\lambda, \tau))] e^{-\lambda^2} d\lambda d\tau \right\|_A$$

$$\leq \frac{1}{\sqrt{\pi}} \sup_{(x,t) \in \bar{D}_\delta} \left\{ \int_0^t \int_{-\infty}^{\infty} |f(v(x + 2\sqrt{t - \tau}\lambda, \tau)) - f(w(x + 2\sqrt{t - \tau}\lambda, \tau))| e^{-\lambda^2} d\lambda d\tau \right\}.$$

Now, as $v, w \in \hat{B}_A^\delta$, then $v(x, t), w(x, t) \in [-\sigma, \sigma]$, $\forall (x, t) \in \bar{D}_\delta$. Thus,

$$|f(v(x, t)) - f(w(x, t))| \le k_0 |v(x, t) - w(x, t)|$$

for all $(x, t) \in \bar{D}_\delta$, via (6.5). Then,

$$\begin{aligned}
&\|G(v) - G(w)\|_A \\
&\le \frac{k_0}{\sqrt{\pi}} \sup_{(x,t) \in \bar{D}_\delta} \left\{ \int_0^t \int_{-\infty}^\infty \left| v \left(x + 2\sqrt{t - \tau}\,\lambda, \tau \right) \right. \right. \\
&\qquad\qquad \left. \left. -w \left(x + 2\sqrt{t - \tau}\,\lambda, \tau \right) \right| e^{-\lambda^2} d\lambda d\tau \right\} \\
&\le \frac{k_0}{\sqrt{\pi}} \|v - w\|_A \int_0^\delta \int_{-\infty}^\infty e^{-\lambda^2} d\lambda d\tau \\
&\le k_0 \delta \|v - w\|_A \\
&\le \frac{1}{2} \|v - w\|_A,
\end{aligned}$$

since $\delta \le \frac{1}{2k_0}$. Therefore,

$$\|G(v) - G(w)\|_A \le \frac{1}{2} \|v - w\|_A$$

for all $v, w \in \hat{B}_A^\delta$. We conclude that $G : \hat{B}_A^\delta \to \hat{B}_A^\delta$ is a contraction mapping, and so the *Banach Fixed Point Theorem* establishes that G has a unique fixed point in $\hat{B}_A^\delta$. That is, there exists a unique $u^* \in \hat{B}_A^\delta$ such that $u^* = G(u^*)$, and so

$$\begin{aligned}
u^*(x, t) &= G(u^*(x, t)) \\
&= \frac{1}{\sqrt{\pi}} \int_{-\infty}^\infty u_0 \left(x + 2\sqrt{t}\lambda \right) e^{-\lambda^2} d\lambda \\
&\quad + \frac{1}{\sqrt{\pi}} \int_0^t \int_{-\infty}^\infty f \left(u^* \left(x + 2\sqrt{t - \tau}\,\lambda, \tau \right) \right) e^{-\lambda^2} d\lambda d\tau; \\
&\forall (x, t) \in \bar{D}_\delta.
\end{aligned}$$

We conclude that the integral equation (a) in Lemma 5.10 has a solution in $\hat{B}_A^\delta$, namely, $u^* \in \hat{B}_A^\delta$. Hence via Lemma 5.10, u^* is a solution of (B-R-D-C) in $\hat{B}_A^\delta$. Thus we have exhibited that (B-R-D-C) has a solution, u^*, on $\bar{D}_\delta$. Uniqueness follows directly from Theorem 6.1. $\qquad\qquad\square$

Example† **6.3** Consider (B-R-D-C) with $f : \mathbb{R} \to \mathbb{R}$ defined by

$$f(u) = u^2(1 - u^3), \quad \forall u \in \mathbb{R}. \qquad (6.8)$$

We observe immediately that $f \in L$. We may conclude from Theorem 6.2 that (B-R-D-C) has a unique solution on $\bar{D}_\delta$, where $\delta = \frac{1}{2k_0}$ since $f(0) = 0$, and k_0 is a Lipschitz constant for f on $[-\sigma, \sigma]$ with $\sigma = 2\|u_0\|_B + 1$. We now calculate a suitable k_0. Observe that f has a bounded derivative on $[-\sigma, \sigma]$ for any $\sigma > 0$:

$$f'(u) = 2u - 5u^4, \quad \forall u \in [-\sigma, \sigma]$$

so that

$$|f'(u)| \leq 2\sigma + 5\sigma^4, \quad \forall u \in [-\sigma, \sigma].$$

Thus,

$$k_0 = 2\sigma + 5\sigma^4 = 2(2\|u_0\|_B + 1) + 5(2\|u_0\|_B + 1)^4$$

provides a suitable Lipschitz constant. We conclude that (B-R-D-C) has a unique solution at least up to $t = \delta$; that is, on $\bar{D}_\delta$ where

$$\delta = \frac{1}{2(2\|u_0\|_B + 1)(2 + 5(2\|u_0\|_B + 1)^3)}. \qquad \lrcorner$$

We next consider how and when a *local* solution to (B-R-D-C) on $\bar{D}_\delta$ may be extended to a *global* solution on $\bar{D}_T$ for given $T > 0$. We have the following theorem.

Theorem† 6.4 (Global Existence) *Let* $f \in L$. *Suppose that (B-R-D-C) is a priori bounded on* $\bar{D}_T$ *for any* $0 \leq T \leq T^*$ *(with bound* l_T). *Then (B-R-D-C) has a unique solution on* $\bar{D}_{T^*}$.

Proof Let k_* be a Lipschitz constant for f on $[-2l_{T^*} - 1, 2l_{T^*} + 1]$. Now put

$$\delta = \frac{1}{2k_* + |f(0)|}.$$

Then, via Theorem 6.2, (B-R-D-C) has a solution u on $\bar{D}_\delta$, since $\|u_0\|_B \leq l_0 \leq l_{T^*}$ (via the a priori bound) and so $[-2\|u_0\|_B - 1, 2\|u_0\|_B + 1] \subseteq [-2l_{T^*} - 1, 2l_{T^*} + 1]$ and hence k_* provides a Lipschitz constant for f on $[-2\|u_0\|_B - 1, 2\|u_0\|_B + 1]$. Since (B-R-D-C) is a priori bounded on $\bar{D}_T$ for each $0 \leq T \leq T^*$, then

$$\|u\|_A \leq l_\delta \text{ on } \bar{D}_\delta,$$

and so

$$\|u(\cdot, \delta)\|_B \leq l_\delta \leq l_{T^*},$$

with, via Lemma 5.12 and Lemma 5.15, $u(\cdot, \delta) \in \mathrm{BPC}^2(\mathbb{R})$.

We can therefore extend u again by applying Theorem 6.2 with $k = k_*$. Thus u is extended onto $\bar{D}_{2\delta}$. By construction, $u : \bar{D}_{2\delta} \to \mathbb{R}$ solves (B-R-D-C), provided we can establish that u_t, u_x, u_{xx} exist and are continuous across $t = \delta$, for all $-\infty < x < \infty$. To this end we observe, via the a priori bound, that

$$\|u(\cdot, \delta/2)\|_B \leq l_{\delta/2} \leq l_{T^*}$$

whilst $u(\cdot, \delta/2) \in \mathrm{BPC}^2(\mathbb{R})$, and we may construct a function $\phi : (-\infty, \infty) \times \left[\frac{\delta}{2}, \frac{3\delta}{2}\right] \to \mathbb{R}$, which solves (B-R-D-C) on $(-\infty, \infty) \times \left[\frac{\delta}{2}, \frac{3\delta}{2}\right]$ with $\phi(x, \frac{\delta}{2}) = u(x, \frac{\delta}{2}) \ \forall \ x \in (-\infty, \infty)$. This follows again from Theorem 6.2. Now ϕ_t, ϕ_x and ϕ_{xx} exist and are continuous in $(-\infty, \infty) \times (\frac{\delta}{2}, \frac{3\delta}{2}]$. Moreover, uniqueness in Theorem 6.1 requires that, $\phi = u$ on $(-\infty, \infty) \times \left[\frac{\delta}{2}, \frac{3\delta}{2}\right]$. It follows that u_t, u_x and u_{xx} exist and are continuous across $t = \delta$.

Thus we have extended the solution to (B-R-D-C), from $\bar{D}_\delta$ onto $\bar{D}_{2\delta}$. Repeated application of this procedure enables us to extend the solution to (B-R-D-C) onto $\bar{D}_{N\delta}$ with $N \in \mathbb{N}$ such that, $(N-1)\delta < T^* \leq N\delta$. Uniqueness follows from Theorem 6.1, and the proof is complete. □

Remark 6.5 Let $f \in L$. When (B-R-D-C) is a priori bounded on $\bar{D}_T$ with $0 \leq T \leq T^*$, for every $T^* > 0$, then (B-R-D-C) has a unique solution on $\bar{D}_\infty$ (with $\bar{D}_\infty = (-\infty, \infty) \times [0, \infty)$). It should be noted that it is often the case when considering initial value problems to refer to global solutions as those which exist on $\bar{D}_\infty$. However, here the convention is inherited from [56]. ⌐

It remains in this section to provide a continuous dependence result for (B-R-D-C) with $f \in L$. Before this, we introduce the following comparison theorem.

Theorem† 6.6 (Lipschitz Comparison) *Let $f \in L$. Furthermore let $\bar{u}$ and $\underline{u}$ be a regular super-solution and a regular sub-solution respectively to (B-R-D-C) on $\bar{D}_T$. Then,*

$$\underline{u}(x, t) \leq \bar{u}(x, t); \quad \forall (x, t) \in \bar{D}_T.$$

Proof Define $w : \bar{D}_T \to \mathbb{R}$, by

$$w(x, t) = \underline{u}(x, t) - \bar{u}(x, t); \quad \forall (x, t) \in \bar{D}_T. \tag{6.9}$$

Then, on D_T, we have via Definition 2.2,

$$w_t - w_{xx} - h(x, t)w \equiv N[\underline{u}] - N[\bar{u}] \leq 0, \tag{6.10}$$

where

$$h(x, t) = \begin{cases} 0 & \text{; when } \bar{u} = \underline{u} \text{ on } \bar{D}_T \\ \dfrac{(f(\underline{u}) - f(\bar{u}))}{(\underline{u} - \bar{u})} & \text{; when } \underline{u} \neq \bar{u} \text{ on } \bar{D}_T. \end{cases}$$

Now as $\underline{u}, \bar{u} : \bar{D}_T \to \mathbb{R}$ are continuous and uniformly bounded as $|x| \to \infty$ for $t \in [0, T]$, then both $\underline{u}$ and $\bar{u}$ are bounded on $\bar{D}_T$, say $|\underline{u}|, |\bar{u}| \leq M$ on $\bar{D}_T$ for some constant $M \geq 0$. It then follows via Definition 2.4 that there exists a constant $k_M > 0$ such that

$$\frac{|f(X) - f(Y)|}{|X - Y|} \leq k_M$$

for all $X, Y \in [-M, M]$ with $X \neq Y$. Thus $h(x, t)$ is bounded above by k_M on $\bar{D}_T$. Furthermore via Definition 2.2,

$$w(x, 0) \leq 0; \quad \forall x \in \mathbb{R}. \tag{6.11}$$

A direct application of Theorem 3.6 with (6.10) and (6.11) establishes that

$$w(x, t) \leq 0; \quad \forall (x, t) \in \bar{D}_T,$$

and via (6.9) we have

$$\underline{u}(x, t) \leq \bar{u}(x, t); \quad \forall (x, t) \in \bar{D}_T,$$

as required. $\square$

We now have the following continuous dependence theorem.

Theorem† **6.7** (Lipschitz Continuous Dependence) *Let* $f \in L'$. *Suppose that* $u_1, u_2 : \bar{D}_T \to \mathbb{R}$ *are solutions to (B-R-D-C) with initial data* $u_{10}, u_{20} \in BPC^2(\mathbb{R})$ *and reaction function* $f : \mathbb{R}^2 \to \mathbb{R}$ *with parameters* α_1 *and* α_2 *respectively, with*

$$\|u_1\|_A, \|u_2\|_A \leq M, \quad |\alpha_1|, |\alpha_2| \leq a.$$

Then,

$$\|u_1(\cdot, t) - u_2(\cdot, t)\|_B \leq (\|u_{10} - u_{20}\|_B + k_A |\alpha_1 - \alpha_2| t) e^{k_U t}; \quad \forall t \in [0, T],$$

where k_U *and* k_A *are the positive constants arising for* $f \in L'$ *when* $U = [-M, M]$ *and* $A = [-a, a]$ *in Definition 2.11.*

Proof Since $(u_i(x, t), \alpha_i) \in U \times A$ for all $(x, t) \in \bar{D}_T$ and $f \in L'$, there exist constants $k_U, k_A > 0$ such that

$$|f(u_1(x, t), \alpha_1) - f(u_2(x, t), \alpha_2)| \leq k_U |u_1(x, t) - u_2(x, t)| + k_A |\alpha_1 - \alpha_2|;$$
$$\forall (x, t) \in \bar{D}_T. \tag{6.12}$$

Now, since u_1 and u_2 are solutions to (B-R-D-C), then, via Lemma 5.10 and (6.12),

$$
\begin{aligned}
&|u_1(x,t) - u_2(x,t)| \\
&\leq \frac{1}{\sqrt{\pi}} \int_{-\infty}^{\infty} \left| u_{10}\left(x + 2\sqrt{t}\lambda\right) - u_{20}\left(x + 2\sqrt{t}\lambda\right) \right| e^{-\lambda^2} d\lambda \\
&\quad + \frac{1}{\sqrt{\pi}} \int_0^t \int_{-\infty}^{\infty} \left| f\left(u_1\left(x + 2\sqrt{t-\tau}\lambda, \tau\right), \alpha_1\right) \right. \\
&\qquad \left. - f\left(u_2\left(x + 2\sqrt{t-\tau}\lambda, \tau\right), \alpha_2\right) \right| e^{-\lambda^2} d\lambda d\tau \\
&\leq \frac{1}{\sqrt{\pi}} \int_{-\infty}^{\infty} \|u_{10} - u_{20}\|_B e^{-\lambda^2} d\lambda \\
&\quad + \frac{1}{\sqrt{\pi}} \int_0^t \int_{-\infty}^{\infty} \left(k_U \left| u_1\left(x + 2\sqrt{t-\tau}\lambda, \tau\right) - u_2\left(x + 2\sqrt{t-\tau}\lambda, \tau\right) \right| \right. \\
&\qquad \left. + k_A |\alpha_1 - \alpha_2| \right) e^{-\lambda^2} d\lambda d\tau \\
&\leq \|u_{10} - u_{20}\|_B + \frac{1}{\sqrt{\pi}} \int_0^t \int_{-\infty}^{\infty} k_U \|u_1(\cdot, \tau) - u_2(\cdot, \tau)\|_B e^{-\lambda^2} d\lambda d\tau \\
&\quad + k_A |\alpha_1 - \alpha_2| t \\
&\leq \|u_{10} - u_{20}\|_B + k_A |\alpha_1 - \alpha_2| t + \int_0^t k_U \|u_1(\cdot, \tau) - u_2(\cdot, \tau)\|_B d\tau;
\end{aligned}
$$

$$\forall (x,t) \in \bar{D}_T. \tag{6.13}$$

Now, since the right hand side of (6.13) is independent of x, upon taking the supremum over all $x \in \mathbb{R}$ on the left hand side, we obtain

$$
\begin{aligned}
\|u_1(\cdot, t) - u_2(\cdot, t)\|_B &\leq \|u_{10} - u_{20}\|_B + k_A |\alpha_1 - \alpha_2| t \\
&\quad + \int_0^t k_U \|u_1(\cdot, \tau) - u_2(\cdot, \tau)\|_B d\tau \quad \forall t \in [0, T].
\end{aligned}
$$

Now since $k_A, k_U > 0$ and $\|u_1(\cdot, t) - u_2(\cdot, t)\|_B$ is continuous for all $t \in [0, T]$ (via Corollary 5.16), via Proposition 5.6, we have

$$\|u_1(\cdot, t) - u_2(\cdot, t)\|_B \leq \left(\|u_{10} - u_{20}\|_B + k_A |\alpha_1 - \alpha_2| t \right) e^{k_U t}; \quad \forall t \in [0, T].$$

This completes the proof. $\qquad\qquad\qquad\qquad\qquad\qquad\qquad\qquad\qquad\square$

Next, we have.

Theorem 6.8 *Let $f \in L$. Suppose that (B-R-D-C) has a solution $u^*: \bar{D}_T \to \mathbb{R}$ when $u_0 = u_0^* \in BPC^2(\mathbb{R})$ (which is unique). Let $M = \|u^*\|_A \geq 0$. Then there* *exists $\delta > 0$ (depending on f, M and T) such that (B-R-D-C) has a solution $u : \bar{D}_T \to \mathbb{R}$ (which is unique) for every $u_0 \in BPC^2(\mathbb{R})$ with*

$$\|u_0 - u_0^*\|_B < \delta$$

and $\|u\|_A \leq \frac{3}{2}M$.

Proof To begin, we introduce the function $\bar{f} \in L$ such that

$$\bar{f}(u) = \begin{cases} f(-(2M+1)) & ; -\infty < u \leq -(2M+1) \\ f(u) & ; -(2M+1) \leq u \leq (2M+1) \\ f((2M+1)) & ; (2M+1) \leq u < \infty \end{cases}$$

and consider (B-R-D-C) with reaction function $\bar{f} \in L$, which we denote as $\overline{\text{(B-R-D-C)}}$. Observe that $u^* : \bar{D}_T \to \mathbb{R}$ is the unique solution of $\overline{\text{(B-R-D-C)}}$ with $u_0 = u_0^* \in BPC^2(\mathbb{R})$. Now take $0 < \delta < 1$, and $u_0 \in BPC^2(\mathbb{R})$ such that

$$\|u_0 - u_0^*\|_B \leq \delta. \tag{6.14}$$

Let $u : \bar{D}_T \to \mathbb{R}$ be any corresponding solution to $\overline{\text{(B-R-D-C)}}$. It follows from the *Lipschitz Comparison* Theorem 6.6 that

$$-(\|u_0\|_B + M't) \leq u(x,t) \leq (\|u_0\|_B + M't); \quad \forall(x,t) \in \bar{D}_T, \tag{6.15}$$

where, M' is given by

$$M' = \sup_{u \in \mathbb{R}} |\bar{f}(u)| = \sup_{|u| \leq 2M+1} |f(u)|.$$

Thus, via (6.14) and (6.15),

$$\|u\|_A \leq (M+1) + M'T, \tag{6.16}$$

and so $\overline{\text{(B-R-D-C)}}$ is a priori bounded on $\bar{D}_T$ and it follows from the *Global Existence* Theorem 6.4, that $\overline{\text{(B-R-D-C)}}$ with $u_0 \in BPC^2(\mathbb{R})$ satisfying (6.14) has a unique global solution on $\bar{D}_T$, say $u : \bar{D}_T \to \mathbb{R}$, and this solution satisfies (6.16), which is independent of δ. It also follows, via (6.16) and (6.14), and the *Lipschitz Continuous Dependence* Theorem 6.7, that

$$\|u(\cdot,t) - u^*(\cdot,t)\|_B \leq \delta e^{k_U t}; \quad \forall t \in [0,T], \tag{6.17}$$

with $k_U > 0$ being a Lipschitz constant for $\bar{f} \in L$ on $[-((M+1) + M'T), ((M+1) + M'T)]$. Note that for any closed bounded interval $E \subset \mathbb{R}$, a Lipschitz constant for f on E will also be a Lipschitz constant for $\bar{f}$ on E. Now, choose

$$\delta = \min\left\{\frac{1}{2}Me^{-k_U T}, 1\right\} \tag{6.18}$$

after which we obtain from (6.17),

$$||u(\cdot,t)||_B \le ||u^*(\cdot,t)||_B + \frac{1}{2}Me^{k_U(t-T)} \le M + \frac{1}{2}M = 3M/2; \quad \forall t \in [0,T],$$

from which we have

$$||u||_A \le 3M/2. \tag{6.19}$$

An immediate consequence of (6.19) is that $u : \bar{D}_T \to \mathbb{R}$ is a solution to (B-R-D-C) on $\bar{D}_T$, and is unique (via *Uniqueness* Theorem 6.1). Thus we have established that for each $u_0 \in BPC^2(\mathbb{R})$ which satisfies (6.14), with δ given by (6.18), then (B-R-D-C) has a unique solution on $\bar{D}_T$, and this solution satisfies (6.19), as required. □

We are now in a position to state the main result of this section which encapsulates the preceding results, namely:

Theorem† 6.9 *Let $f \in L$. Suppose (B-R-D-C) is a priori bounded on $\bar{D}_T$ for all $T \ge 0$ and for all $u_0 \in BPC^2(\mathbb{R})$. Then (B-R-D-C) is globally well-posed on $BPC^2(\mathbb{R})$.*

Proof Since (B-R-D-C) is a priori bounded on $\bar{D}_T$ for each $T > 0$ and for each $u_0 \in BPC^2(\mathbb{R})$, it follows from the Global Existence Theorem 6.4, that (B-R-D-C) has a unique solution $u : \bar{D}_T \to \mathbb{R}$ for each $T > 0$ and for each $u_0 \in BPC^2(\mathbb{R})$. Hence (P1) is satisfied. This solution is unique, via the Uniqueness Theorem 6.1 (since $f \in L$), and so (P2) is satisfied. Now let $u_0^* \in BPC^2(\mathbb{R})$ with corresponding solution $u^* : \bar{D}_T \to \mathbb{R}$. It follows from Theorem 6.8 that there exists $\triangle > 0$ such that for all $u_0 \in BPC^2(\mathbb{R})$ with $||u_0 - u_0^*||_B < \triangle$, then, $||u||_A \le 3M/2$, with $M = ||u^*||_A$, and so $||u||_A, ||u^*||_A \le 3M/2$. An application of the Lipschitz Continuous Dependence Theorem 6.7 then gives

$$||u(\cdot,t) - u^*(\cdot,t)||_B \le ||u_0 - u_0^*||_B e^{k_U t} < \triangle e^{k_U T}; \quad \forall t \in [0,T], \tag{6.20}$$

with k_U being a Lipschitz constant for $f \in L$ on $[-3M/2, 3M/2]$. Thus, given any $\epsilon > 0$, take

$$\delta = \min\left\{\triangle, \epsilon e^{-k_U T}\right\} \tag{6.21}$$

after which, for each $u_0 \in BPC^2(\mathbb{R})$ satisfying

$$||u_0 - u_0^*||_B < \delta,$$

we have via (6.21) and (6.20),

$$||u(\cdot,t) - u^*(\cdot,t)||_B < \epsilon; \quad \forall t \in [0,T],$$

and so

$$||u - u^*||_A < \epsilon,$$

and (P3) is satisfied. The proof is complete.   □

As a final note, we give a condition on solutions of (B-R-D-C) which cannot be continued beyond a finite time.

Theorem† 6.10 (Blow-up)
Let $f \in L$ and $u_0 \in BPC^2(\mathbb{R})$. Let $u : \bar{D}_{T^*} \backslash (\mathbb{R} \times \{T^*\}) \to \mathbb{R}$ be a solution to (B-R-D-C) which cannot be continued onto $\bar{D}_{T^*}$. Then $||u(\cdot, t)||_B$ is unbounded as $t \to T^{*-}$.

Proof Suppose that $||u(\cdot, t)||_B$ is bounded as $t \to T^{*-}$. Then $||u(\cdot, t)||_B$ is bounded for $t \in [0, T^*)$. Hence there exists $M > 0$ such that

$$||u(\cdot, t)||_B \leq M; \quad \forall t \in [0, T^*).$$

Since $f \in L$, there exists a Lipschitz constant $k_M > 0$ for f on $[-\sigma, \sigma]$ where $\sigma = 2M + 1$. Now consider (B-R-D-C) with initial data $u_0^* : \mathbb{R} \to \mathbb{R}$ where

$$u_0^*(x) = u\left(x, T^* - \delta/2\right); \quad \forall x \in \mathbb{R},$$

$$\delta = \frac{1}{2k_M + |f(0)|}.$$

It follows from Lemma 5.12 and Lemma 5.15 that $u_0^* \in BPC^2(\mathbb{R})$ and so, via Theorem 6.2, that there exists a function $u^* : \bar{D}_\delta \to \mathbb{R}$ that uniquely solves (B-R-D-C) with initial data $u_0^* : \mathbb{R} \to \mathbb{R}$. It follows that

$$u^*(x, t) = u\left(x, T^* - \delta/2 + t\right); \quad \forall (x, t) \in \mathbb{R} \times \left[0, \delta/2\right).$$

Therefore (as in the proof of Theorem 6.4) the function $u_c : \bar{D}_{T^*+\delta/2} \to \mathbb{R}$, given by

$$u_c(x, t) = \begin{cases} u(x, t) & ; (x, t) \in \bar{D}_{T^*} \backslash (\mathbb{R} \times \{T^*\}) \\ u^*(x, t) & ; (x, t) \in \mathbb{R} \times \left[T^*, T^* + \delta/2\right] \end{cases}$$

solves (B-R-D-C) on $\bar{D}_{T^*+\delta/2}$, and is a continuation of u onto $\bar{D}_{T^*+\delta/2}$, and we arrive at a contradiction. We conclude that $||u(\cdot, t)||_B$ must be unbounded as $t \to T^{*-}$.   □

Remark 6.11 The theory developed within this chapter is an adaptation, for a more general class of solutions and reaction functions, of the theory contained

in [56]. Specifically, in [56], a solution $u : \bar{D}_T \to \mathbb{R}$ to (B-R-D-C) satisfies
Definition 2.1 with the additional condition

$$u(x, t) \to 0 \text{ as } |x| \to \infty \text{ uniformly for } t \in [0, T]$$

whilst the reaction function $f \in L$ satisfies the corresponding condition
$f(0) = 0$. These restrictions have been removed in this monograph.   ⌐

7

The Bounded Reaction-Diffusion Cauchy Problem with $f \in L_u$

This chapter follows a similar format to the previous chapter, but the extent of results is not as broad as in Chapter 6. This is principally due to the lack of a generic existence result for (B-R-D-C) with $f \in L_u$. However, a number of significant results relating to the well-posedness of (B-R-D-C) have been obtained. Throughout we take $u_0 \in \mathrm{BPC}^2(\mathbb{R})$. We begin with a comparison theorem.

Theorem‡ 7.1 (Comparison) *Let $f \in L_u$. Furthermore let $\bar{u}$ and $\underline{u}$ be a regular super-solution and a regular sub-solution respectively to (B-R-D-C) on $\bar{D}_T$. Then,*

$$\underline{u}(x,t) \leq \bar{u}(x,t); \quad \forall (x,t) \in \bar{D}_T.$$

Proof Define $w : \bar{D}_T \to \mathbb{R}$, by

$$w(x,t) = \underline{u}(x,t) - \bar{u}(x,t); \quad \forall (x,t) \in \bar{D}_T. \tag{7.1}$$

Then, on D_T, we have via Definition 2.2,

$$w_t - w_{xx} - h(x,t)w \equiv N[\underline{u}] - N[\bar{u}] \leq 0, \tag{7.2}$$

where

$$h(x,t) = \begin{cases} 0 & ; \text{ when } \bar{u} = \underline{u} \text{ on } \bar{D}_T \\ \frac{(f(\underline{u}) - f(\bar{u}))}{(\underline{u} - \bar{u})} & ; \text{ when } \underline{u} \neq \bar{u} \text{ on } \bar{D}_T. \end{cases} \tag{7.3}$$

Now as $\underline{u}, \bar{u} : \bar{D}_T \to \mathbb{R}$ are continuous and uniformly bounded as $|x| \to \infty$ for $t \in [0, T]$, then both $\underline{u}$ and $\bar{u}$ are bounded on $\bar{D}_T$, say $|\underline{u}|, |\bar{u}| \leq M$ on $\bar{D}_T$ for some constant $M \geq 0$. It then follows via Proposition 2.8 that there exists a constant $k_M > 0$ such that

$$\frac{f(X) - f(Y)}{X - Y} \leq k_M$$

71

for all $X, Y \in [-M, M]$ with $X \neq Y$. Thus $h(x, t)$ is bounded above by k_M on $\bar{D}_T$. Furthermore, via Definition 2.2,

$$w(x, 0) \leq 0; \quad \forall x \in \mathbb{R}. \tag{7.4}$$

A direct application of Theorem 3.6 with (7.2) and (7.4) establishes that

$$w(x, t) \leq 0; \quad \forall (x, t) \in \bar{D}_T,$$

and via (7.1) we have

$$\underline{u}(x, t) \leq \overline{u}(x, t); \quad \forall (x, t) \in \bar{D}_T,$$

as required.   □

It should be observed that a strong version of Theorem 7.1 is unobtainable as Example 3.4 illustrates. Moreover, observe that the proof of Theorem 7.1 differs from that of Theorem 6.6 only in that an upper-Lipschitz constant is sufficient to show that h given by (7.3) is bounded above. We are now able to establish uniqueness for (B-R-D-C) when $f \in L_u$.

Theorem$_\ddagger$ 7.2 (Uniqueness for $f \in L_u$) *Let $f \in L_u$, then (B-R-D-C) has at most one solution on $\bar{D}_T$ for any $T > 0$.*

Proof Let $u^{(1)} : \bar{D}_T \to \mathbb{R}$ and $u^{(2)} : \bar{D}_T \to \mathbb{R}$ both be solutions to (B-R-D-C) on $\bar{D}_T$. It is trivial to show that if u is a solution to (B-R-D-C) on $\bar{D}_T$ then, via Definition 2.2, u is both a (R-S-P) and a (R-S-B) to (B-R-D-C) on $\bar{D}_T$. On taking $u^{(1)}$ and $u^{(2)}$ to be a (R-S-B) and (R-S-P) respectively, then via Theorem 7.1 we have

$$u^{(1)} \leq u^{(2)} \text{ on } \bar{D}_T.$$

By a symmetrical argument, we have

$$u^{(2)} \leq u^{(1)} \text{ on } \bar{D}_T.$$

Therefore,

$$u^{(1)} = u^{(2)} \text{ on } \bar{D}_T,$$

as required.   □

Remark 7.3 Although we have established that (B-R-D-C) has at most one solution on $\bar{D}_T$ when $f \in L_u$, it is yet to be established whether such a solution exists. This question remains open at present.   ⌐

We next establish a conditional continuous dependence result. We have,

Theorem‡ 7.4 *Let $f \in L_u'$, and let $u_1, u_2 : \bar{D}_T \rightarrow \mathbb{R}$ be (unique) solutions to (B-R-D-C) corresponding to $u_0 = u_0^1 : \mathbb{R} \rightarrow \mathbb{R}$ and $u_0 = u_0^2 : \mathbb{R} \rightarrow \mathbb{R}$, where $u_0^1, u_0^2 \in BPC^2(\mathbb{R})$, and $\alpha = \alpha_1$ and $\alpha = \alpha_2$, respectively. Let M_U and M_A be positive constants such that*

$$\max\{\|u_1\|_A, \|u_2\|_A\} \le M_U, \quad \max\{|\alpha_1|, |\alpha_2|\} \le M_A.$$

Suppose further that $f = f(u, \alpha)$ is non-decreasing with respect to $\alpha \in [-M_A, M_A]$ for each $u \in [-M_U, M_U]$, and

$$\alpha_2 \ge \alpha_1,$$

$$u_0^2(x) - u_0^1(x) \ge 0; \quad \forall x \in \mathbb{R}, \tag{7.5}$$

then

$$\|u_2(\cdot, t) - u_1(\cdot, t)\|_B \le \left(\|u_0^2 - u_0^1\|_B + k_A(\alpha_2 - \alpha_1)t\right)e^{k_U t}; \quad \forall t \in [0, T],$$

where $k_A > 0$ is a Lipschitz constant for $f(u, \alpha)$ with respect to $\alpha \in [-M_A, M_A]$ uniformly for $u \in [-M_U, M_U]$, and, k_U is an upper Lipschitz constant for $f(u, \alpha)$ with respect to $u \in [-M_U, M_U]$ uniformly for $\alpha \in [-M_A, M_A]$.

Proof Under the above conditions on $f(u, \alpha)$ for $(u, \alpha) \in [-M_U, M_U] \times [-M_A, M_A]$, it is straightforward to verify that $u_1 : \bar{D}_T \rightarrow \mathbb{R}$ is a regular sub-solution and $u_2 : \bar{D}_T \rightarrow \mathbb{R}$ is a regular super-solution to that (B-R-D-C) with $\alpha = \alpha_1$ and $u_0 = u_0^1$. It then follows from Comparison Theorem 7.1 (since $f(\cdot, \alpha) \in L_u$ for $\alpha \in \{\alpha_1, \alpha_2\}$) that

$$u_1(x, t) \le u_2(x, t); \quad \forall (x, t) \in \bar{D}_T.$$

Now, via the conditions on $f(u, \alpha)$ and Lemma 5.11 we have

$$0 \le (u_2 - u_1)(x, t)$$
$$\le \|u_0^2 - u_0^1\|_B + \frac{1}{\sqrt{\pi}} \int_0^t \int_{-\infty}^{\infty} (f(u_2, \alpha_2)$$
$$- f(u_1, \alpha_1)) \left(x + 2\sqrt{t - \tau}\lambda, \tau\right) e^{-\lambda^2} d\lambda d\tau$$
$$= \|u_0^2 - u_0^1\|_B + \frac{1}{\sqrt{\pi}} \int_0^t \int_{-\infty}^{\infty} (f(u_2, \alpha_2) - f(u_1, \alpha_2)) \tag{7.6}$$
$$\left(x + 2\sqrt{t - \tau}\lambda, \tau\right) e^{-\lambda^2} d\lambda d\tau$$
$$+ \frac{1}{\sqrt{\pi}} \int_0^t \int_{-\infty}^{\infty} (f(u_1, \alpha_2) - f(u_1, \alpha_1)) \left(x + 2\sqrt{t - \tau}\lambda, \tau\right) e^{-\lambda^2} d\lambda d\tau$$

$$\leq \|u_0^2 - u_0^1\|_B + \frac{1}{\sqrt{\pi}} \int_0^t \int_{-\infty}^{\infty} k_U(u_2 - u_1)$$

$$(x + 2\sqrt{t - \tau}\lambda, \tau)\, e^{-\lambda^2} d\lambda d\tau + k_A(\alpha_2 - \alpha_1)t$$

$$\leq \|u_0^2 - u_0^1\|_B + k_A(\alpha_2 - \alpha_1)t$$

$$+ k_U \int_0^t \|(u_2 - u_1)(\cdot, \tau)\|_B d\tau \quad \forall (x, t) \in \bar{D}_T. \tag{7.7}$$

Since the right hand side of (7.7) is independent of x, then we have

$$\|(u_2 - u_1)(\cdot, t)\|_B \leq \|u_0^2 - u_0^1\|_B + k_A(\alpha_2 - \alpha_1)t$$

$$+ k_U \int_0^t \|(u_2 - u_1)(\cdot, \tau)\|_B d\tau; \quad \forall t \in [0, T]. \tag{7.8}$$

As $\|(u_2 - u_1)(\cdot, t)\|_B \in L^1([0, T])$ (via Lemma 5.5), an application of Proposition 5.6 to (7.8), gives

$$\|(u_2 - u_1)(\cdot, t)\|_B \leq \left(\|u_0^2 - u_0^1\|_B + k_A(\alpha_2 - \alpha_1)t\right) e^{k_U t}; \quad \forall t \in [0, T], \tag{7.9}$$

as required. □

A corollary to this result, which removes the ordering on the initial data, is,

Corollary‡ 7.5 *Let $f \in L_u'$ and satisfy all of the conditions given in Theorem 7.4. Let $u_1 : \bar{D}_T \to \mathbb{R}$ and $u_2 : \bar{D}_T \to \mathbb{R}$ be as described in Theorem 7.4 with the exception of condition (7.5). In addition, let $u_3 : \bar{D}_T \to \mathbb{R}$ be a (unique) solution to (B-R-D-C) corresponding to $u_0 = u_0^3 : \mathbb{R} \to \mathbb{R}$ and $\alpha = \alpha_2$. Let M_A and M_U be positive constants such that*

$$\max\{\|u_1\|_A, \|u_2\|_A, \|u_3\|_A\} \leq M_U, \quad \max\{|\alpha_1|, |\alpha_2|\} \leq M_A.$$

Suppose that $\alpha_2 \geq \alpha_1$ and for $i = 1, 2$

$$\|u_0^3 - u_0^i\|_B \leq \delta \quad and \quad u_0^3(x) \geq u_0^i(x); \quad \forall x \in \mathbb{R}$$

with $\delta \geq 0$. Then,

$$\max_{i,j=1,2,3} \|u_i(\cdot, t) - u_j(\cdot, t)\|_B \leq (2\delta + t k_A(\alpha_2 - \alpha_1)) e^{k_U t}; \quad \forall t \in [0, T],$$

where $k_A > 0$ and $k_U > 0$ are as defined in Theorem 7.4.

Proof We may apply Theorem 7.4 to obtain

$$\|u_3(\cdot, t) - u_1(\cdot, t)\|_B \leq \left(\|u_0^3 - u_0^1\|_B + t k_A(\alpha_2 - \alpha_1)\right) e^{k_U t}; \quad \forall t \in [0, T], \tag{7.10}$$

$$\|u_3(\cdot, t) - u_2(\cdot, t)\|_B \leq \left(\|u_0^3 - u_0^2\|_B\right) e^{k_U t}; \quad \forall t \in [0, T]. \quad (7.11)$$

Now,

$$\|u_2(\cdot, t) - u_1(\cdot, t)\|_B \leq \|u_3(\cdot, t) - u_1(\cdot, t)\|_B + \|u_3(\cdot, t) - u_2(\cdot, t)\|_B$$
$$\leq \left(\|u_0^3 - u_0^1\|_B + \|u_0^3 - u_0^2\|_B\right) e^{k_U t}$$
$$+ t k_A (\alpha_2 - \alpha_1) e^{k_U t} \quad (7.12)$$

for all $t \in [0, T]$ via (7.10), (7.11) and the triangle inequality. However,

$$\max\{\|u_0^3 - u_0^1\|_B, \|u_0^3 - u_0^2\|_B\} \leq \delta$$

and so, it follows from (7.12) that

$$\|u_2(\cdot, t) - u_1(\cdot, t)\|_B \leq (2\delta + t k_A (\alpha_2 - \alpha_1)) e^{k_U t}, \quad \forall t \in [0, T]. \quad (7.13)$$

The result follows from (7.10), (7.11) and (7.13). $\qquad\qquad\square$

We now have,

Theorem‡ 7.6 *Let $f \in L_u$ and suppose that (B-R-D-C) has a (unique) solution $u : \bar{D}_T \to \mathbb{R}$ for every $u_0 \in BPC^2(\mathbb{R})$. Let $u_0^* \in BPC^2(\mathbb{R})$ have the corresponding (unique) solution $u^* : \bar{D}_T \to \mathbb{R}$. Then given any $\epsilon > 0$, and any $u_0 \in BPC^2(\mathbb{R})$ such that*

$$\|u_0 - u_0^*\|_B < \min\left\{\frac{1}{2}, \frac{1}{3}\epsilon e^{-k_U T}\right\},$$

it follows that

$$\|u - u^*\|_A < \epsilon.$$

Here $k_U > 0$ is an upper Lipschitz constant for $f \in L_u$ on the interval $[-M_U, M_U]$, with $M_U > 0$ depending upon u_0^* and T.

Proof Let $u^* : \bar{D}_T \to \mathbb{R}$ be the (unique) solution to (B-R-D-C) with $u_0 = u_0^* \in BPC^2(\mathbb{R})$, and $u_\delta^* : \bar{D}_T \to \mathbb{R}$ be the (unique) solution to (B-R-D-C) with $u_0 = u_0^* + \delta \in BPC^2(\mathbb{R})$, with $0 < \delta \leq 1/2$. In addition, let $u^{\pm} : \bar{D}_T \to \mathbb{R}$ be the unique solutions to (B-R-D-C) with $u_0 = \inf_{x \in \mathbb{R}} u_0^*(x) - 1 \in BPC^2(\mathbb{R})$ and $u_0 = \sup_{x \in \mathbb{R}} u_0^*(x) + 1 \in BPC^2(\mathbb{R})$, respectively. It follows from Theorem 7.2 and the translation invariance of the reaction-diffusion equation in (B-R-D-C), that there exist $U_+, U_- \in C^1([0, T])$ such that $u^+(x, t) = U_+(t)$ and $u^-(x, t) = U_-(t)$ for all $(x, t) \in \bar{D}_T$. Now let $u_0 \in BPC^2(\mathbb{R})$ such that

$$\|u_0 - u_0^*\|_B < \delta, \quad (7.14)$$

with corresponding solution $u : \bar{D}_T \to \mathbb{R}$. It then follows from the Comparison Theorem 7.1 with (7.14), that

$$U_-(t) \leq u^*(x, t) \leq U_+(t),$$

$$U_-(t) \leq u_\delta^*(x, t) \leq U_+(t), \tag{7.15}$$

$$U_-(t) \leq u(x, t) \leq U_+(t),$$

for all $(x, t) \in \bar{D}_T$. Thus,

$$||u^*||_A, \ ||u_\delta^*||_A, \ ||u||_A \leq M_U, \tag{7.16}$$

where $M_U > 0$ is given by

$$M_U = \max \left\{ \sup_{t \in [0,T]} |U_-(t)|, \ \sup_{t \in [0,T]} |U_+(t)| \right\}.$$

An application of Theorem 7.4 now gives

$$||u_\delta^*(\cdot, t) - u^*(\cdot, t)||_B \leq \delta e^{k_U t},$$

$$||u(\cdot, t) - u_\delta^*(\cdot, t)||_B \leq 2\delta e^{k_U t}, \tag{7.17}$$

for all $t \in [0, T]$ with $k_U > 0$ being an upper Lipschitz constant for $f \in L_u$ on $[-M_U, M_U]$. It follows from (7.17) and the triangle inequality that

$$||u(\cdot, t) - u^*(\cdot, t)||_B \leq ||u(\cdot, t) - u_\delta^*(\cdot, t)||_B$$
$$+ ||u_\delta^*(\cdot, t) - u^*(\cdot, t)||_B \leq 3\delta e^{k_U t}; \quad \forall t \in [0, T]. \tag{7.18}$$

Now set $\delta = \min \left\{ \frac{1}{2}, \frac{1}{3} \epsilon e^{-k_U T} \right\}$ and the result follows from (7.14) and (7.18).

$\square$

Corollary$_\ddagger$ 7.7 *Under the conditions of Theorem 7.6, with $BPC^2(\mathbb{R})$ replaced by $BPC_+^2(\mathbb{R})$ throughout, then the same conclusion holds. Similarly, for any closed interval $I \subset \mathbb{R}$, $BPC^2(\mathbb{R})$ may be replaced by either of*

$$A_I(\mathbb{R}) = \left\{ u_0 \in BPC^2(\mathbb{R}) : u_0(x) \in I; \ \forall x \in \mathbb{R} \right\},$$

$$A_{I+}(\mathbb{R}) = \left\{ u_0 \in BPC_+^2(\mathbb{R}) : u_0(x) \in I; \ \forall x \in \mathbb{R} \right\},$$

with the same conclusion holding in Theorem 7.6.

Proof For $\text{BPC}_+^2(\mathbb{R})$, the proof follows the same steps as the proof of Theorem 7.6 upon replacing the initial data for u^- : $\bar{D}_T \to \mathbb{R}$ from $u_0 = \inf_{x \in \mathbb{R}}\{u_0^*(x)\} - 1$ to $u_0 = 0$ for all $x \in \mathbb{R}$, since the former is not guaranteed to be in the set $\text{BPC}_+^2(\mathbb{R})$. The proof is similar for $A_I(\mathbb{R})$ and $A_{I+}(\mathbb{R})$.   □

We now have the following conditional global well-posedness result.

Corollary‡ 7.8 *Let $f \in L_u$ and suppose that (B-R-D-C) has a solution $u : \bar{D}_T \to \mathbb{R}$ for every $u_0 \in BPC^2(\mathbb{R})$ and any $T > 0$. Then (B-R-D-C) is globally well-posed on $BPC^2(\mathbb{R})$. An equivalent statement holds with $BPC^2(\mathbb{R})$ replaced by $BPC_+^2(\mathbb{R})$, $A_I(\mathbb{R})$ or $A_{I+}(\mathbb{R})$.*

Proof For any of the initial data sets concerned, (P1) is satisfied according to the conditions of the corollary and (P2) follows from Theorem 7.2. For $\text{BPC}^2(\mathbb{R})$, (P3) follows from Theorem 7.6 and for $\text{BPC}_+^2(\mathbb{R})$, $A_I(\mathbb{R})$ and $A_{I+}(\mathbb{R})$, (P3) follows from Corollary 7.7. The proof is complete.   □

With an additional technical condition on solutions of (B-R-D-C) with $f \in L_u$, we can improve Corollary 7.8 to obtain a conditional uniform global well-posedness result, namely,

Theorem‡ 7.9 *Let $f \in L_u$ and suppose that (B-R-D-C) has a (unique) solution $u : \bar{D}_\infty \to \mathbb{R}$ for every $u_0 \in BPC^2(\mathbb{R})$. Let $u_0^* \in BPC^2(\mathbb{R})$ have the corresponding (unique) solution $u^* : \bar{D}_\infty \to \mathbb{R}$. Moreover, suppose that there exists $T' \geq 0$, such that for any $u_0 \in BPC^2(\mathbb{R})$, the corresponding solution $u : \bar{D}_\infty \to \mathbb{R}$ satisfies*

$$u(x, t) \in E \subset \mathbb{R}; \quad \forall (x, t) \in \bar{D}_\infty^{T'},$$

with E being a closed bounded interval and where $f \in L_u$ is non-increasing on E. Then given any $\epsilon > 0$, there exists $\delta > 0$, depending only upon T', u_0^*, f and ϵ, such that for any $u_0 \in BPC^2(\mathbb{R})$ that satisfies

$$||u_0 - u_0^*||_B < \delta,$$

it follows that for any $T > 0$,

$$||(u - u^*)(\cdot, t)||_B < \epsilon; \quad \forall t \in [0, T].$$

Proof Without loss of generality let $T' \geq 1$. Let $u^* : \bar{D}_\infty \to \mathbb{R}$ be the unique solution to (B-R-D-C) with $u_0 = u_0^* \in \text{BPC}^2(\mathbb{R})$. In addition let $u_\pm^* : \bar{D}_{T'} \to \mathbb{R}$ be the unique solutions to (B-R-D-C) with $u_0 = \inf_{x \in \mathbb{R}} u_0^*(x) - 1 \in \text{BPC}^2(\mathbb{R})$

and $u_0 = \sup_{x \in \mathbb{R}} u_0^*(x) + 1 \in \text{BPC}^2(\mathbb{R})$, respectively. It follows from Theorem 7.2 and the translational invariance of the reaction-diffusion equation in (B-R-D-C), that there exist $U_+, U_- \in C^1([0, T'])$ such that $u_+^*(x, t) = U_+(t)$ and $u_-^*(x, t) = U_-(t)$ for all $(x, t) \in \bar{D}_{T'}$. Now let

$$M_U = \sup_{t \in [0, T']} \{\max\{|U_-(t)|, |U_+(t)|\}\}$$

and set $k_U > 0$ to be an upper Lipschitz constant for $f \in L_u$ on $[-M_U, M_U]$. Then, given $\epsilon > 0$, via Theorem 7.6, there exists $\delta' > 0$, depending on T', u_0^*, f and ϵ, such that, for all $u_0 \in \text{BPC}^2(\mathbb{R})$, with corresponding solution $u : \bar{D}_\infty \to \mathbb{R}$, which satisfy $\|u_0 - u_0^*\|_B \le \delta'$, we have

$$\|(u - u^*)(\cdot, t)\|_B < \frac{\epsilon}{4(1 + 2T'k_U)}; \quad \forall t \in [0, T']. \tag{7.19}$$

Now, set

$$\delta = \min\left\{\delta', \frac{1}{8}\epsilon, 1\right\} \tag{7.20}$$

and henceforth consider $u_0 \in \text{BPC}^2(\mathbb{R})$ such that $\|u_0 - u_0^*\|_B < \delta$. Next, let $u_\delta^* : \bar{D}_\infty \to \mathbb{R}$ be the unique solution to (B-R-D-C) with initial data $u_0 = u_0^* + \delta \in \text{BPC}^2(\mathbb{R})$. Then, via Theorem 7.1,

$$\max\{u^*(x, t), u(x, t)\} \le u_\delta^*(x, t); \quad \forall(x, t) \in \bar{D}_\infty. \tag{7.21}$$

Thus, it follows from (7.21) and Lemma 5.11 that

$$0 \le u_\delta^*(x, t) - u(x, t)$$

$$= \frac{1}{\sqrt{\pi}} \int_{-\infty}^{\infty} \left(u_0^*\left(x + 2\sqrt{t}\lambda\right) + \delta - u_0\left(x + 2\sqrt{t}\lambda\right)\right) e^{-\lambda^2} d\lambda$$

$$+ \frac{1}{\sqrt{\pi}} \int_0^t \int_{-\infty}^{\infty} \left(f\left(u_\delta^*\left(x + 2\sqrt{t - \tau}\lambda, \tau\right)\right)\right.$$

$$\left. - f\left(u\left(x + 2\sqrt{t - \tau}\lambda, \tau\right)\right)\right) e^{-\lambda^2} d\lambda d\tau$$

$$\le 2\delta + \frac{1}{\sqrt{\pi}} \int_0^{T'} \int_{-\infty}^{\infty} \left(f\left(u_\delta^*\left(x + 2\sqrt{t - \tau}\lambda, \tau\right)\right)\right.$$

$$\left. - f\left(u\left(x + 2\sqrt{t - \tau}\lambda, \tau\right)\right)\right) e^{-\lambda^2} d\lambda d\tau$$

$$+ \frac{1}{\sqrt{\pi}} \int_{T'}^t \int_{-\infty}^{\infty} \left(f\left(u_\delta^*\left(x + 2\sqrt{t - \tau}\lambda, \tau\right)\right)\right.$$

$$\left. - f\left(u\left(x + 2\sqrt{t - \tau}\lambda, \tau\right)\right)\right) e^{-\lambda^2} d\lambda d\tau \tag{7.22}$$

for all $(x, t) \in \bar{D}_T^{T'}$ and any $T > T'$. In addition, it follows from (7.20) and Theorem 7.1 that

$$U_-(t) \leq u^*(x,t) \leq U_+(t),$$

$$U_-(t) \leq u_\delta^*(x,t) \leq U_+(t),$$

$$U_-(t) \leq u(x,t) \leq U_+(t)$$

for all $(x,t) \in \bar{D}_{T'}$. Thus, for $B_A^{T'}$, we conclude that

$$\max\{\|u^*\|_A, \|u_\delta^*\|_A, \|u\|_A\} \leq M_U. \tag{7.23}$$

Therefore, it follows from (7.21), (7.19) and Lemma 5.5 that

$$\frac{1}{\sqrt{\pi}} \int_0^{T'} \int_{-\infty}^{\infty} \left(f\left(u_\delta^*\left(x + 2\sqrt{t-\tau}\lambda, \tau\right)\right) \right.$$
$$\left. - f\left(u\left(x + 2\sqrt{t-\tau}\lambda, \tau\right)\right)\right) e^{-\lambda^2} d\lambda d\tau$$

$$\leq \frac{1}{\sqrt{\pi}} \int_0^{T'} \int_{-\infty}^{\infty} k_U \left(u_\delta^*\left(x + 2\sqrt{t-\tau}\lambda, \tau\right)\right.$$
$$\left. - u\left(x + 2\sqrt{t-\tau}\lambda, \tau\right)\right) e^{-\lambda^2} d\lambda d\tau$$

$$\leq \int_0^{T'} k_U \|(u_\delta^* - u)(\cdot, \tau)\|_B d\tau$$

$$< \int_0^{T'} \frac{2k_U \epsilon}{4(1 + 2T'k_U)} d\tau$$

$$< \frac{1}{4}\epsilon \tag{7.24}$$

for all $(x,t) \in \bar{D}_T^{T'}$. Additionally, since $u_\delta^*(x,t), u(x,t) \in E$ for all $(x,t) \in \bar{D}_T^{T'}$, then it follows that

$$\frac{1}{\sqrt{\pi}} \int_{T'}^{t} \int_{-\infty}^{\infty} \left(f\left(u_\delta^*\left(x + 2\sqrt{t-\tau}\lambda, \tau\right)\right) \right.$$
$$\left. - f\left(u\left(x + 2\sqrt{t-\tau}\lambda, \tau\right)\right)\right) e^{-\lambda^2} d\lambda d\tau \leq 0 \tag{7.25}$$

for all $(x,t) \in \bar{D}_T^{T'}$, via (7.21) and observing that $f \in L_u$ is non-increasing on E. Thus, it follows from (7.22), (7.24), (7.25) and (7.20) that

$$0 \leq u_\delta^*(x,t) - u(x,t) < 2\delta + \frac{1}{4}\epsilon \leq \frac{1}{4}\epsilon + \frac{1}{4}\epsilon = \frac{1}{2}\epsilon \tag{7.26}$$

for all $(x,t) \in \bar{D}_T^{T'}$. Since the right hand side of (7.26) is independent of x, then we have

$$\|(u_\delta^* - u)(\cdot, t)\|_B < \frac{1}{2}\epsilon \tag{7.27}$$

for all $t \in [T', T]$. Moreover, since (7.27) holds for any $T \geq T'$, it follows that (7.27) holds for $t \in [T', \infty)$. Thus, we conclude from (7.20), (7.19) and (7.27) that

$$||(u_\delta^* - u)(\cdot, t)||_B < \frac{\epsilon}{2} \qquad (7.28)$$

for all $t \in [0, \infty)$. Thus, it follows from (7.28) that

$$||(u^* - u)(\cdot, t)||_B \leq ||(u^* - u_\delta^*)(\cdot, t)||_B + ||(u_\delta^* - u)(\cdot, t)||_B < \frac{\epsilon}{2} + \frac{\epsilon}{2} = \epsilon$$

for all $t \in [0, \infty)$. The result then follows for δ given by (7.20), as required. $\qquad \square$

Corollary$_\ddagger$ 7.10 *In Theorem 7.9, the initial data set $BPC^2(\mathbb{R})$ can be replaced by either $BPC_+^2(\mathbb{R})$, $A_I(\mathbb{R})$ or $A_{I+}(\mathbb{R})$ with the same conclusion holding.*

Proof For $BPC_+^2(\mathbb{R})$, the result follows on replacing $BPC^2(\mathbb{R})$ by $BPC_+^2(\mathbb{R})$ in the proof of Theorem 7.9. The proof is similar for $A_I(\mathbb{R})$ and $A_{I+}(\mathbb{R})$. $\qquad \square$

Corollary$_\ddagger$ 7.11 *Let $f \in L_u$ and suppose that (B-R-D-C) has a solution $u : \bar{D}_\infty \to \mathbb{R}$ for every $u_0 \in BPC^2(\mathbb{R})$. Moreover, suppose that there exists a $T' \geq 0$, such that for any $u_0 \in BPC^2(\mathbb{R})$, the corresponding solution $u : \bar{D}_\infty \to \mathbb{R}$ satisfies*

$$u(x, t) \in E \subset \mathbb{R}; \quad \forall (x, t) \in \bar{D}_\infty^{T'},$$

with E being a closed bounded interval and where $f \in L_u$ is non-increasing on E. Then, (B-R-D-C) is uniformly globally well-posed on $BPC^2(\mathbb{R})$. An equivalent statement holds with $BPC^2(\mathbb{R})$ replaced by $BPC_+^2(\mathbb{R})$, $A_I(\mathbb{R})$ or $A_{I+}(\mathbb{R})$.

Proof For any $u_0 \in BPC^2(\mathbb{R})$, $BPC_+^2(\mathbb{R})$, $A_I(\mathbb{R})$ and $A_{I+}(\mathbb{R})$, (P1) is satisfied according to the conditions of the corollary and (P2) follows from Theorem 7.2. For $BPC^2(\mathbb{R})$, (P3) follows from Theorem 7.9 and for $BPC_+^2(\mathbb{R})$, $A_I(\mathbb{R})$ and $A_{I+}(\mathbb{R})$, (P3) follows from Corollary 7.10. The proof is complete. $\qquad \square$

Example 7.12 Consider the (B-R-D-C) with reaction function $f \in L_u$, given by

$$f(u) = [u^p]^+ (u - 1/2) [(1 - u)^q]^+; \quad \forall u \in \mathbb{R},$$

with $p, q \in (0, 1)$. We can immediately state,

(i) Suppose (B-R-D-C) has a solution $u : \bar{D}_\infty \to \mathbb{R}$ for all $u_0 \in \text{BPC}^2(\mathbb{R})$. Then (B-R-D-C) is globally well-posed on $\text{BPC}^2(\mathbb{R})$ via Corollary 7.8.

(ii) Suppose (B-R-D-C) has a solution $u : \bar{D}_\infty \to \mathbb{R}$ for all $u_0 \in A_I(\mathbb{R})$ with I being a closed bounded interval such that $I \subset (-\infty, 1/2)$ or $I \subset (1/2, \infty)$. Then (B-R-D-C) is uniformly globally well-posed on $A_I(\mathbb{R})$ via Corollary 7.11 upon taking $E = [a, u_{min}]$, where $a = \min\{0, \min I\}$ and $u_{min} \in (0, 1/2)$ with $f(u_{min}) = \inf_{u \in \mathbb{R}} f(u)$, and $E = [u_{max}, b]$, where $u_{max} \in (1/2, 1)$ with $f(u_{max}) = \sup_{u \in \mathbb{R}} f(u)$ and $b = \max\{1, \max I\}$, respectively.   ⌐

Remark 7.13 The development of the theory in this chapter was motivated by the observation in [36] that a related problem to (B-R-D-C) has uniqueness for $f \in L_u$ together with an associated comparison theorem, which suggests that development of the theory when $f \in L_u$ would be fruitful. It should be noted that non-increasing functions $f \in L_u$ have been considered in related problems (see Theorem 5, [21] (p.201)). For additional development of the theory in this chapter see [50].   ⌐

8

The Bounded Reaction-Diffusion Cauchy Problem with $f \in H_\alpha$

In this chapter, we develop an approach to establishing a local existence result for $f \in H_\alpha$ with global existence obtained under the condition of a priori bounds. This approach is a significant generalisation of that considered in [5] and [54] for specific cases of non-Lipschitz continuous nonlinearities. However, unlike the corresponding result for $f \in L$ (Theorem 6.2), uniqueness is not obtained for $f \in H_\alpha$. Additionally, from the construction of the local existence result for $f \in H_\alpha$, a conditional comparison theorem is obtained for $f \in H_\alpha$. It should be noted that theory similar to what follows has been developed in [61] and [14] using an alternative approach, relating principally to the Dirichlet problem on compact spatial domains. Once existence has been established, we obtain qualitative information about the structure of maximal and minimal solutions. With these qualitative results, we then establish a conditional continuous dependence-type result.

Definition 8.1 Let $f \in H_\alpha$ for some $\alpha \in (0, 1)$ and $u_0 \in \mathrm{BPC}^2(\mathbb{R})$. Let

$$S = \left\{ u : \bar{D}_T \to \mathbb{R} : u \text{ is a solution to the given (B-R-D-C) on } \bar{D}_T \right\}.$$

Then $\bar{u} : \bar{D}_T \to \mathbb{R}$ is said to be a *maximal solution* to the given (B-R-D-C) when $\bar{u} \in S$ and for all $u \in S$,

$$\bar{u}(x, t) \geq u(x, t); \quad \forall (x, t) \in \bar{D}_T.$$

Correspondingly, $\underline{u} : \bar{D}_T \to \mathbb{R}$ is said to be a *minimal solution* to the given (B-R-D-C) when $\underline{u} \in S$ and for all $u \in S$,

$$\underline{u}(x, t) \leq u(x, t); \quad \forall (x, t) \in \bar{D}_T. \qquad \lrcorner$$

Remark 8.2 For a given (B-R-D-C), when $\underline{u} = \bar{u}$ on $\bar{D}_T$, then (B-R-D-C) has a unique solution on $\bar{D}_T$. $\qquad \lrcorner$

We now state one of the main results of this chapter.

Theorem‡ 8.3 (Local Hölder Existence) *Consider (B-R-D-C) with $f \in H_\alpha$ for some $\alpha \in (0, 1)$, and $u_0 \in BPC^2(\mathbb{R})$. Then there exist a minimal and a maximal solution to (B-R-D-C) on $\bar{D}_\delta$, with*

$$\delta = \min\left\{ \frac{(m_0 + a')}{c'}, \frac{(m_0 - b')}{c'} \right\} \geq \frac{1}{2c'},$$

where $m_0 = \|u_0\|_B + 1$, $a' = \inf_{x \in \mathbb{R}} u_0(x) - 1/2$, $b' = \sup_{x \in \mathbb{R}} u_0(x) + 1/2$ and

$$c' = \max\left\{ \left| \inf_{y \in [-m_0, m_0]}\{f(y)\} - 1 \right|, \left| \sup_{y \in [-m_0, m_0]}\{f(y)\} + 1 \right| \right\}.$$

In addition, with $\underline{u} : \bar{D}_\delta \to \mathbb{R}$ and $\bar{u} : \bar{D}_\delta \to \mathbb{R}$ being the minimal and maximum solutions respectively, then

$$\max\left\{ \|\underline{u}\|_A, \|\bar{u}\|_A \right\} \leq m_0. \qquad \lrcorner$$

In what follows we develop a constructional proof of Theorem 8.3, and, in doing so, we establish Proposition 8.17. As a consequence of this we have the following remark concerning Theorem 8.3:

Remark 8.4 Let $\bar{u}, \underline{u} : \bar{D}_\delta \to \mathbb{R}$ be the maximal and minimal solutions to (B-R-D-C) as given in Theorem 8.3. Then $\bar{u}$ and $\underline{u}$ are, respectively, maximal and minimal solutions to (B-R-D-C) on $\bar{D}_{\delta'}$, for any $0 < \delta' \leq \delta$, and on $\bar{D}_{\delta_2}^{\delta_1}$, for any $0 \leq \delta_1 < \delta_2 \leq \delta$. Now, let $\bar{u}^c : \bar{D}_T \to \mathbb{R}$ be a function obtained by repeated application of Theorem 8.3 and glueing together the associated maximal solution and its domain at each stage. Then $\bar{u}^c : \bar{D}_T \to \mathbb{R}$ is a solution to (B-R-D-C), and is a maximal solution to (B-R-D-C) on $\bar{D}_T$. Similarly let $\underline{u}^c : \bar{D}_T \to \mathbb{R}$ be a function obtained by repeated application of Theorem 8.3 and glueing together the associated minimal solution and its domain at each stage. Then $\underline{u}^c : \bar{D}_T \to \mathbb{R}$ is a solution to (B-R-D-C), and is a minimal solution to (B-R-D-C) on $\bar{D}_T$. In what follows, we will refer to $\bar{u}^c : \bar{D}_T \to \mathbb{R}$ (when it exists on $\bar{D}_T$) as a *constructed maximal solution* to (B-R-D-C) on $\bar{D}_T$. Similarly, we will refer to $\underline{u}^c : \bar{D}_T \to \mathbb{R}$ (when it exists on $\bar{D}_T$) as a *constructed minimal solution* to (B-R-D-C) on $\bar{D}_T$.

Note that a constructed maximal (minimal) solution to (B-R-D-C) on $\bar{D}_T$ is a maximal (minimal) solution to (B-R-D-C) on $\bar{D}_T$. However the converse does not necessarily follow; a maximal (minimal) solution to (B-R-D-C) on $\bar{D}_T$ need not be a constructed maximal (minimal) solution to (B-R-D-C) on $\bar{D}_T$. $\lrcorner$

Immediate consequences of the above are,

Corollary‡ **8.5** *Let* $f \in H_\alpha$ *for some* $\alpha \in (0, 1)$ *and* $u_0 \in BPC^2(\mathbb{R})$. *Then there exists a global constructed maximal (minimal) solution to (B-R-D-C) on* $\bar{D}_\infty$, *or there exists* T_u $(T_l) > 0$ *such that (B-R-D-C) has a constructed maximal (minimal) solution on* $\bar{D}_{T_u} \backslash (\mathbb{R} \times \{T_u\})$ ($\bar{D}_{T_l} \backslash (\mathbb{R} \times \{T_l\})$) *which cannot be continued onto* $\bar{D}_{T_u}$ ($\bar{D}_{T_l}$).

Proof This follows directly from repeated application of Theorem 8.3 to (B-R-D-C) and Remark 8.4.   □

Corollary‡ **8.6** *Let* $f \in H_\alpha$ *for some* $\alpha \in (0, 1)$ *and* $u_0 \in BPC^2(\mathbb{R})$. *Let* $\bar{u}^c(\underline{u}^c) : \bar{D}_{T^*} \backslash (\mathbb{R} \times \{T^*\}) \to \mathbb{R}$ *be a constructed maximal (minimal) solution to (B-R-D-C) which cannot be continued onto* $\bar{D}_{T^*}$. *Then* $\|\bar{u}^c(\cdot, t)\|_B$ $(\|\underline{u}^c(\cdot, t)\|_B)$ *is unbounded as* $t \to T^{*-}$.

Proof This follows similar steps to the proof of Theorem 6.10, via Theorem 8.3 and Remark 8.4.   □

To begin to establish Theorem 8.3, we must first prove a denseness result, namely,

Proposition‡ **8.7** (Lipschitz Density) *Consider* $f \in H_\alpha$ *with* $\alpha \in (0, 1)$. *Let* $k_H > 0$ *be a Hölder constant for* f *on the closed bounded interval* $E \subset \mathbb{R}$. *Then, on* E, *given any* $\epsilon > 0$, *there exists a Lipschitz continuous function* $g : E \to \mathbb{R}$ *such that*

$$|f(x) - g(x)| < \epsilon; \quad \forall x \in E,$$

where g *is also a Hölder continuous function of degree* α *on* E *with Hölder constant* $3k_H$.

Proof Let $E \subset \mathbb{R}$ be a closed bounded interval, and $k_H > 0$ be a Hölder constant for f on E. Now, given any $\epsilon > 0$, set δ as follows,

$$\delta = \left(\frac{\epsilon}{2k_H} \right)^{1/\alpha}. \tag{8.1}$$

Then, for all $x, y \in E$, with $|x - y| < \delta$, we have

$$|f(y) - f(x)| < \frac{\epsilon}{2}. \tag{8.2}$$

We may write $E = [a, b] \subset \mathbb{R}$. Now take $N \in \mathbb{N}$ with $N > \frac{(b-a)}{\delta}$ and divide the interval E into uniform sub-intervals X_n ($n = 1, \ldots, N$), defined by

$$X_n = [x_{n-1}, x_n], \text{ where } x_0 = a, \; x_N = b, \; x_n = x_{n-1} + \frac{(b-a)}{N} \qquad (8.3)$$

for each $1 \le n \le N$. Next define $l_n : X_n \to \mathbb{R}$, for $1 \le n \le N$, as

$$l_n(x) = \left(\frac{f(x_n)(x - x_{n-1}) + f(x_{n-1})(x_n - x)}{x_n - x_{n-1}} \right); \quad \forall x \in X_n \qquad (8.4)$$

and define $g : E \to \mathbb{R}$ such that on each interval $X_n \subset E$,

$$g(x) = l_n(x); \quad \forall x \in X_n.$$

Note that g defined by (8.3) and (8.4) is Lipschitz continuous on E with Lipschitz constant given by

$$k_E^l = \max_{1 \le n \le N} \left| \frac{f(x_n) - f(x_{n-1})}{(x_n - x_{n-1})} \right|.$$

Let $x \in E$, then there exists n such that $x \in X_n$ for some $n = 1, 2, \ldots, N$ and so

$$|f(x) - g(x)| \le |f(x) - f(x_n)| + |f(x_n) - g(x)|$$
$$= |f(x) - f(x_n)| + |g(x_n) - g(x)| < \frac{\epsilon}{2} + \frac{\epsilon}{2} = \epsilon,$$

via (8.1), (8.2), (8.3) and (8.4). It remains to shown that g is also Hölder continuous of degree $0 < \alpha < 1$ on E with Hölder constant $3k_H$. Observe that since $g(x_n) = f(x_n)$ for each $n = 0, 1, 2, \ldots, N$, then on each interval X_n, we have

$$\left| \frac{dg}{dx} \right| = \left| \frac{f(x_n) - f(x_{n-1})}{x_n - x_{n-1}} \right| \le \left| \frac{(x_n - x_{n-1})^\alpha k_H}{x_n - x_{n-1}} \right|$$
$$= |x_n - x_{n-1}|^{\alpha-1} k_H \quad \forall x \in X_n. \qquad (8.5)$$

It follows from the mean value theorem with (8.5), that for any $x, y \in X_n$,

$$|g(x) - g(y)| \le |x_n - x_{n-1}|^{\alpha-1} k_H |x - y|$$
$$= k_H \left| \frac{x - y}{x_n - x_{n-1}} \right|^{1-\alpha} |x - y|^\alpha \le k_H |x - y|^\alpha. \qquad (8.6)$$

Now for $x \in X_n$ and $y \in X_m$ where, without loss of generality, $m > n$, then via (8.6) and (8.4),

$$|g(x) - g(y)| \le |g(x) - g(x_n)| + |f(x_n) - f(x_{m-1})| + |g(x_{m-1}) - g(y)|$$
$$\le k_H |x - x_n|^\alpha + k_H |x_n - x_{m-1}|^\alpha + k_H |x_{m-1} - y|^\alpha$$
$$\le 3k_H |x - y|^\alpha, \qquad (8.7)$$

since $x \leq x_n \leq x_{m-1} \leq y$. Inequalities (8.6) and (8.7) establish that g is Hölder continuous of degree α on E with Hölder constant $3k_H$, as required. $\square$

Remark 8.8 Let $f \in H_\alpha$ for some $\alpha \in (0, 1)$ and $E = [a, b]$ for $b > a$. If $g : E \to \mathbb{R}$ is constructed as in Proposition 8.7, then $f(a) = g(a)$ and $f(b) = g(b)$. $\lrcorner$

Next we proceed to construct two sequences of functions which will later be shown to converge to the minimal and maximal solutions to (B-R-D-C).

Proposition$_\ddagger$ 8.9 *Let $f \in H_\alpha$ for some $\alpha \in (0, 1)$, and $E = [a, b]$ be a closed bounded interval. Let $k_H > 0$ be a Hölder constant for f on $[a, b]$. Then there exist sequences $\{\overline{f}_n\}_{n \in \mathbb{N}}$ and $\{\underline{f}_n\}_{n \in \mathbb{N}}$, such that for each $n \in \mathbb{N}$ the functions $\overline{f}_n, \underline{f}_n : \mathbb{R} \to \mathbb{R}$ satisfy.*

(a) $\overline{f}_n$ and $\underline{f}_n$ are Lipschitz continuous on every closed bounded interval $E' \subset \mathbb{R}$.

(b) $\overline{f}_n$ and $\underline{f}_n$ are Hölder continuous of degree α on every closed bounded interval $E' \subset \mathbb{R}$, with Hölder constant independent of $n \in \mathbb{N}$.

(c) $\overline{f}_n(u) \to f(u)$ and $\underline{f}_n(u) \to f(u)$ as $n \to \infty$ uniformly for all $u \in E$.

(d) $\underline{f}_n(u) \leq f(u) \leq \overline{f}_n(u)$ for all $u \in E$ and for each $n \in \mathbb{N}$.

(e) $\overline{f}_{n+1}(u) \leq \overline{f}_n(u)$ and $\underline{f}_{n+1}(u) \geq \underline{f}_n(u)$ for all $u \in \mathbb{R}$ and for each $n \in \mathbb{N}$.

Proof The Lipschitz Denseness result in Proposition 8.7 guarantees that there exists a sequence of Lipschitz continuous functions $g_n : [a, b] \to \mathbb{R}$ (each of which is also Hölder continuous on $[a, b]$ of degree $0 < \alpha < 1$, with Hölder constant $3k_H$ on $[a, b]$) such that $g_n(a) = f(a)$, $g_n(b) = f(b)$ and which satisfy

$$\sup_{u \in E}\{|f - g_n|(u)\} \leq 1/2^n, \tag{8.8}$$

for each $n \in \mathbb{N}$. Now define $\overline{f}_n, \underline{f}_n : \mathbb{R} \to \mathbb{R}$, for each $n \in \mathbb{N}$, to be

$$\overline{f}_n(u) = \begin{cases} g_{n+2}(u) + \frac{1}{2^n} & ; u \in [a, b] \\ g_{n+2}(a) + \frac{1}{2^n} & ; u \in (-\infty, a) \\ g_{n+2}(b) + \frac{1}{2^n} & ; u \in (b, \infty), \end{cases} \quad \underline{f}_n(u) = \begin{cases} g_{n+2}(u) - \frac{1}{2^n} & ; u \in [a, b] \\ g_{n+2}(a) - \frac{1}{2^n} & ; u \in (-\infty, a) \\ g_{n+2}(b) - \frac{1}{2^n} & ; u \in (b, \infty). \end{cases} \tag{8.9}$$

We now give the proof for $\{\underline{f}_n\}$, with the proof for $\{\overline{f}_n\}$ following similarly. Statements (a) and (b) follow immediately from (8.9). Next we observe that

$$|\underline{f}_n - f|(u) \le |g_{n+2} - f|(u) + 1/2^n \le 5/2^{n+2}, \tag{8.10}$$

for all $u \in [a, b]$ and $n \in \mathbb{N}$, and so $\underline{f}_n(u) \to f(u)$ as $n \to \infty$ uniformly for $u \in [a, b]$, which establishes statement (c). Also observe that for any $u \in [a, b]$ and $n \in \mathbb{N}$, we have

$$\underline{f}_n(u) = g_{n+2}(u) - 1/2^n \le (f(u) + 1/2^{n+2}) - 1/2^n \le f(u) - 3/2^{n+2} \le f(u), \tag{8.11}$$

from which statement (d) follows. It remains to establish that the sequence $\{\underline{f}_n\}_{n \in \mathbb{N}}$ is non-decreasing on $\mathbb{R}$. Observe via (8.8) and (8.9), that for any $n \in \mathbb{N}$,

$$\underline{f}_{n+1}(u) \ge \left(f(u) - \frac{1}{2^{n+3}} \right) - \frac{1}{2^{n+1}} = f(u) - \frac{5}{2^{n+3}}, \tag{8.12}$$

$$\underline{f}_n(u) \le \left(f(u) + \frac{1}{2^{n+2}} \right) - \frac{1}{2^n} = f(u) - \frac{6}{2^{n+3}}, \tag{8.13}$$

for all $u \in [a, b]$. Combining (8.12) and (8.13) gives

$$\underline{f}_{n+1}(u) - \underline{f}_n(u) \ge \frac{1}{2^{n+3}} > 0 \tag{8.14}$$

for all $u \in [a, b]$. In addition it follows from (8.9) that

$$\underline{f}_{n+1}(u) - \underline{f}_n(u) = \frac{1}{2^{n+1}} > 0, \tag{8.15}$$

for all $u \in (-\infty, a) \cup (b, \infty)$. Statement (e) follows from (8.14) and (8.15). This completes the proof for $\{\underline{f}_n\}$. $\square$

Remark 8.10 In developing the proof of Theorem 8.3, for the given $f \in H_\alpha$ and $u_0 \in \mathrm{BPC}^2(\mathbb{R})$ associated with (B-R-D-C), we will use the corresponding sequences $\{\underline{f}_n\}_{n \in \mathbb{N}}$ and $\{\overline{f}_n\}_{n \in \mathbb{N}}$ as constructed in Proposition 8.9, with the interval $[a, b] = [-m_0, m_0]$ where $m_0 = \|u_0\|_B + 1$.   ⌟

We now consider the sequences of (B-R-D-C) problems with reaction functions $f = \overline{f}_n$ and $f = \underline{f}_n$ as in (8.9), and initial data $(u_0 + 1/(2n)) \in \mathrm{BPC}^2(\mathbb{R})$ and $(u_0 - 1/(2n)) \in \mathrm{BPC}^2(\mathbb{R})$, respectively. Henceforth, these sequences of problems will be referred to as (B-R-D-C)$_n^u$ and (B-R-D-C)$_n^l$ respectively, for each $n \in \mathbb{N}$ (here superscripts u and l indicate upper and lower respectively). We now investigate the problems (B-R-D-C)$_n^u$ and (B-R-D-C)$_n^l$.

Proposition‡ 8.11 *For each $n \in \mathbb{N}$, any solution $\overline{u}_n, \underline{u}_n : \bar{D}_T \to \mathbb{R}$ to the problems $(B\text{-}R\text{-}D\text{-}C)_n^u$ and $(B\text{-}R\text{-}D\text{-}C)_n^l$ respectively, satisfy the inequalities*

$$-c't + a' \leq \underline{u}_n(x,t) \leq \overline{u}_n(x,t) \leq c't + b',$$

for all $(x,t) \in \bar{D}_T$, and any $T > 0$, where

$$c' = \max \left\{ \left| \inf_{y \in [-m_0, m_0]} \{f(y)\} - 1 \right|, \left| \sup_{y \in [-m_0, m_0]} \{f(y)\} + 1 \right| \right\},$$

$$a' = \inf_{x \in \mathbb{R}} u_0(x) - \frac{1}{2}, \quad b' = \sup_{x \in \mathbb{R}} u_0(x) + \frac{1}{2}.$$

Proof For convenience, we define $\overline{v}, \underline{v} : \bar{D}_T \to \mathbb{R}$ to be

$$\underline{v}(x,t) = a' - c't,$$

$$\overline{v}(x,t) = b' + c't,$$

for all $(x,t) \in \bar{D}_T$. We now make a straightforward application of Theorem 7.1, in which we take $\underline{v}$ and $\underline{u}_n$, $\underline{u}_n$ and $\overline{u}_n$, and $\overline{u}_n$ and $\overline{v}$ as regular subsolutions and regular supersolutions to $(B\text{-}R\text{-}D\text{-}C)_n^l$, $(B\text{-}R\text{-}D\text{-}C)_n^u$ and $(B\text{-}R\text{-}D\text{-}C)_n^u$ respectively, which follows on observing

$$\underline{v}_t - \underline{v}_{xx} + c' \leq 0, \quad \underline{u}_{nt} - \underline{u}_{nxx} + c' = \underline{f}_n(\underline{u}_n) + c' \geq 0, \qquad (8.16)$$

$$\underline{u}_{nt} - \underline{u}_{nxx} - \overline{f}_n(\underline{u}_n) = \underline{f}_n(\underline{u}_n) - \overline{f}_n(\underline{u}_n) \leq 0, \quad \overline{u}_{nt} - \overline{u}_{nxx} - \overline{f}_n(\overline{u}_n) = 0 \geq 0, \qquad (8.17)$$

$$\overline{u}_{nt} - \overline{u}_{nxx} - c' = \overline{f}_n(\overline{u}_n) - c' \leq 0, \quad \overline{v}_t - \overline{v}_{xx} - c' \geq 0, \qquad (8.18)$$

on D_T, whilst,

$$\underline{v}(x,0) \leq \underline{u}_n(x,0) < \overline{u}_n(x,0) \leq \overline{v}(x,0), \qquad (8.19)$$

for all $x \in \mathbb{R}$. Now applying Theorem 7.1 to each previously stated pair of regular subsolutions and regular supersolutions gives

$$a' - c't \leq \underline{u}_n(x,t) \leq \overline{u}_n(x,t) \leq b' + c't,$$

for all $(x,t) \in \bar{D}_T$, as required. $\qquad \square$

Remark 8.12 Proposition 8.11 ensures that, with $\delta > 0$ as given in Theorem 8.3,

$$-m_0 \leq a' - c't \leq \underline{u}_n(x,t) \leq \overline{u}_n(x,t) \leq b' + c't \leq m_0$$

for all $(x, t) \in \bar{D}_\delta$. Hence (B-R-D-C)$_n^u$ and (B-R-D-C)$_n^l$ are a priori bounded on $\bar{D}_\delta$, for each $n \in \mathbb{N}$, with a priori bounds independent of $n \in \mathbb{N}$.   ⌐

Proposition‡ 8.13 *The problems (B-R-D-C)$_n^u$ and (B-R-D-C)$_n^l$ ($n \in \mathbb{N}$) have unique solutions $\bar{u}_n : \bar{D}_\delta \to \mathbb{R}$ and $\underline{u}_n : \bar{D}_\delta \to \mathbb{R}$ respectively. Moreover the inequalities in Proposition 8.11 and Remark 8.12 hold on $\bar{D}_\delta$.*

Proof It follows from Remark 8.12 that each of (B-R-D-C)$_n^l$ and (B-R-D-C)$_n^u$ is a priori bounded on $\bar{D}_\delta$ for each $n \in \mathbb{N}$. Furthermore, Proposition 8.9 ensures $\overline{f}_n, \underline{f}_n \in L$ for each $n \in \mathbb{N}$. It then follows from Theorem 6.4 that (B-R-D-C)$_n^u$ and (B-R-D-C)$_n^l$ have unique solutions on $\bar{D}_\delta$ for each $n \in \mathbb{N}$. These solutions must satisfy the inequalities in Proposition 8.11 and Remark 8.12 on $\bar{D}_\delta$.   □

Now that both of the sequences of functions $\{\underline{u}_n\}_{n \in \mathbb{N}}$ and $\{\bar{u}_n\}_{n \in \mathbb{N}}$ have been constructed, it remains to show that they converge to the respective minimal and maximal solutions of the original (B-R-D-C). The remainder of the theory will be presented only for the minimal solution with the theory for the maximal solution following exactly the same steps. We next establish derivative estimates on $\underline{u}_n : \bar{D}_\delta \to \mathbb{R}$. In particular,

Proposition‡ 8.14 *Let $\underline{u}_n : \bar{D}_\delta \to \mathbb{R}$ be the (unique) solution to (B-R-D-C)$_n^l$ ($n \in \mathbb{N}$). Then, on D_δ, we have*

$$|\underline{u}_{nx}(x, t)| \leq \frac{2c'}{\sqrt{\pi}}(1 + \delta^{1/2}) + M_0',$$

$$|\underline{u}_{nt}(x, t)| \leq \frac{2^{(\alpha+1)} I_\alpha}{\alpha \sqrt{\pi}} k_\delta (1 + \delta^{\alpha/2}) + c' + M_0''$$

for all $(x, t) \in D_\delta$. Here, $k_H > 0$ is a Hölder constant for $f \in H_\alpha$ on $[-m_0, m_0]$, and

$$M_0' = \sup_{x \in \mathbb{R}} |u_0'(x)|,$$

$$M_0'' = \sup_{x \in \mathbb{R}} |u_0''(x)|,$$

$$k_\delta = 3k_H \left(\frac{2c'}{\sqrt{\pi}}(1 + \delta^{1/2}) + M_0' \right)^\alpha,$$

$$I_\alpha = \int_{-\infty}^{\infty} |\lambda|^\alpha \left| \lambda^2 - 1/2 \right| e^{-\lambda^2} d\lambda.$$

Proof This follows directly from Lemma 5.12 and Lemma 5.15, on recalling that $\underline{f}_n : \mathbb{R} \to \mathbb{R}$ is Hölder continuous of degree α on $[-m_0, m_0] \subset \mathbb{R}$, with Hölder constant $3k_H$, $\underline{u}'_n(\cdot, 0) = u'_0$ and $\underline{u}''_n(\cdot, 0) = u''_0$ for all $n \in \mathbb{N}$. $\qquad\square$

Remark 8.15 We observe that all bounds in Proposition 8.14 are independent of $n \in \mathbb{N}$. $\qquad\lrcorner$

Before examining the limit of the sequence $\{\underline{u}_n\}_{n \in \mathbb{N}}$, two further results are required. The first is used to show that the sequence $\{\underline{u}_n\}_{n \in \mathbb{N}}$ is non-decreasing. The second is used to exhibit part of a comparison theorem. This can be achieved similarly for the sequence $\{\overline{u}\}_{n \in \mathbb{N}}$.

Proposition$_\ddagger$ 8.16 *Let $\underline{u}_n$, $\underline{u}_{n+1} : \bar{D}_\delta \to \mathbb{R}$ be the unique solutions to (B-R-D-C)^l_n and (B-R-D-C)$^l_{n+1}$ respectively. Then for each $n \in \mathbb{N}$,*

$$\underline{u}_{n+1}(x, t) \geq \underline{u}_n(x, t); \quad \forall\, (x, t) \in \bar{D}_\delta.$$

Proof Recall from Proposition 8.9 that $\underline{f}_n : \mathbb{R} \to \mathbb{R}$ is such that $f \in L$ for any $n \in \mathbb{N}$, and

$$\underline{f}_{n+1}(u) \geq \underline{f}_n(u)$$

for all $u \in \mathbb{R}$. Since $\underline{u}_{n+1}(\cdot, 0) > \underline{u}_n(\cdot, 0)$, the result then follows via a simple application of Theorem 7.1. $\qquad\square$

Proposition$_\ddagger$ 8.17 *Let $\underline{u}_n : \bar{D}_\delta \to \mathbb{R}$ be the unique solution to (B-R-D-C)^l_n on $\bar{D}_\delta$ and $v : \bar{D}_\delta \to \mathbb{R}$ be continuous, bounded and have continuous derivatives v_t, v_x and v_{xx} on D_δ, and such that*

$$v_t - v_{xx} - f(v) \geq 0$$

for all $(x, t) \in D_\delta$. Suppose in addition, that

$$v(x, 0) \geq u_0(x)$$

for all $x \in \mathbb{R}$. Then for all $(x, t) \in \bar{D}_\delta$,

$$\underline{u}_n(x, t) \leq v(x, t).$$

Proof To begin fix $n \in \mathbb{N}$. Since v is bounded on $\bar{D}_\delta$, there exists $M > 0$ such that

$$|v(x, t)| \leq M; \quad \forall (x, t) \in \bar{D}_\delta.$$

When $M \leq m_0$, then

$$v_t - v_{xx} - \underline{f}_n(v) \geq f(v) - \underline{f}_n(v) \geq 0,$$

$$\underline{u}_{nt} - \underline{u}_{nxx} - \underline{f}_n(\underline{u}_n) = 0 \leq 0$$

for all $(x, t) \in D_\delta$, via Proposition 8.9, whilst

$$v(x, 0) \geq u_0(x) > \underline{u}_n(x, 0); \quad \forall x \in \mathbb{R}. \tag{8.20}$$

Upon taking v and $\underline{u}_n$ as a regular supersolution and regular subsolution respectively, an application of Theorem 7.1 gives

$$\underline{u}_n(x, t) \leq v(x, t); \quad \forall (x, t) \in \bar{D}_\delta. \tag{8.21}$$

When $M > m_0$ define $\underline{f}'_n : \mathbb{R} \to \mathbb{R}$ by

$$\underline{f}'_n(u) = \begin{cases} \underline{f}_n(u) & ; u \in [-m_0, m_0] \\ g_{n+2}^-(u) - 1/2^n & ; u \in [-M, -m_0] \\ g_{n+2}^+(u) - 1/2^n & ; u \in [m_0, M] \\ g_{n+2}^-(-M) - 1/2^n & ; u \in (-\infty, -M) \\ g_{n+2}^+(M) - 1/2^n & ; u \in (M, \infty), \end{cases} \tag{8.22}$$

where $g_n^- : [-M, -m_0] \to \mathbb{R}$ and $g_n^+ : [m_0, M] \to \mathbb{R}$ are constructed as in Proposition 8.7, and hence are Lipschitz continuous on $[-M, -m_0]$ and $[m_0, M]$ respectively, and

$$\max \left\{ \sup_{u \in [-M, -m_0]} |g_n^-(u) - f(u)|, \sup_{u \in [m_0, M]} |g_n^+(u) - f(u)| \right\} < 1/2^n.$$

Moreover, via Remark 8.8 and arguments contained in the proof of Proposition 8.9, $\underline{f}'_n \in L$ and $\underline{f}'_n(u) \leq f(u)$ for all $u \in [-M, M]$. Now, taking v and $\underline{u}_n$ to be a regular supersolution and regular subsolution respectively, which follows from (8.20) and the inequalities

$$v_t - v_{xx} - \underline{f}'_n(v) \geq f(v) - \underline{f}'_n(v) \geq 0,$$

$$\underline{u}_{nt} - \underline{u}_{nxx} - \underline{f}'_n(\underline{u}) = \underline{f}_n(\underline{u}_n) - \underline{f}'_n(\underline{u}_n) = 0 \leq 0$$

for all $(x, t) \in D_\delta$, we apply Theorem 7.1 to v and $\underline{u}_n$ which gives

$$\underline{u}_n(x, t) \leq v(x, t); \quad \forall (x, t) \in \bar{D}_\delta. \tag{8.23}$$

The result follows from (8.21) and (8.23), as required. $\square$

Remark 8.18 Note that in Proposition 8.17, any solution $u : \bar{D}_\delta \to \mathbb{R}$ to (B-R-D-C) on $\bar{D}_\delta$ satisfies the conditions on v. Therefore, for all $n \in \mathbb{N}$,

$$\underline{u}_n(x, t) \le u(x, t),$$

for all $(x, t) \in \bar{D}_\delta$. ⌐

Proposition 8.17 and Remark 8.18 guarantee that any limit function of $\{\underline{u}_n\}_{n\in\mathbb{N}}$ is less than or equal to any solution of (B-R-D-C) on $\bar{D}_\delta$. Therefore, if a limit function of $\{\underline{u}_n\}_{n\in\mathbb{N}}$ is itself a solution to (B-R-D-C), then it must be a minimal solution. We now proceed to establish that the sequence $\{\underline{u}_n\}_{n\in\mathbb{N}}$ does indeed have a limit in $\|\cdot\|_A$, and that the limit function provides a solution to (B-R-D-C) on $\bar{D}_\delta$.

For each $(x, t) \in \bar{D}_\delta$, we consider the real sequence

$$\left\{\underline{u}_n(x, t)\right\}_{n\in\mathbb{N}}. \tag{8.24}$$

It follows from Proposition 8.16 and Remark 8.12, that this real sequence is non-decreasing and bounded above, and hence is convergent to, say, $u_l(x, t)$, so that

$$\underline{u}_n(x, t) \to u_l(x, t) \text{ as } n \to \infty \text{ for each } (x, t) \in \bar{D}_\delta. \tag{8.25}$$

It follows, moreover, from Remark 8.12 that

$$-m_0 \le u_l(x, t) \le m_0 \text{ for each } (x, t) \in \bar{D}_\delta. \tag{8.26}$$

Thus we may introduce the function $u^* : \bar{D}_\delta \to \mathbb{R}$ given by

$$u^*(x, t) = u_l(x, t); \quad \forall\, (x, t) \in \bar{D}_\delta, \tag{8.27}$$

and we have from (8.25) that

$$\underline{u}_n \to u^* \text{ as } n \to \infty \text{ pointwise on } \bar{D}_\delta. \tag{8.28}$$

We also have from (8.26), that

$$-m_0 \le u^*(x, t) \le m_0; \quad \forall\, (x, t) \in \bar{D}_\delta. \tag{8.29}$$

Next we have,

Lemma‡ 8.19 *The sequence of functions $\{\underline{u}_n\}_{n\in\mathbb{N}}$ has a subsequence $\{\underline{u}_{n_j}\}_{j\in\mathbb{N}}$ (with $1 \le n_1 < n_2 < n_3 < \ldots$ and $n_j \to \infty$ as $j \to \infty$) such that*

$$\underline{u}_{n_j} \to u^* \text{ as } j \to \infty \text{ uniformly on } \bar{D}_\delta^{0,X},$$

for every $X > 0$. Moreover $u^* : \bar{D}_\delta \to \mathbb{R}$ is continuous on $\bar{D}_\delta$.

Proof Consider the sequence of functions $\{\underline{u}_n\}_{n \in \mathbb{N}}$ in $\bar{D}_\delta$. Then each function $\underline{u}_n$ for $n \in \mathbb{N}$, is continuous on $\bar{D}_\delta$ as it is a solution to (B-R-D-C)$_n^l$ on $\bar{D}_\delta$. Also, we have, for each $n \in \mathbb{N}$,

$$|\underline{u}_n(x,t)| \leq m_0; \quad \forall (x,t) \in \bar{D}_\delta, \tag{8.30}$$

via Remark 8.12. Now $\underline{u}_{nt}$ and $\underline{u}_{nx}$ exist and are continuous on D_δ and so it follows from the mean value theorem that for any (x_0, t_0), $(x_1, t_1) \in D_\delta$, then

$$|\underline{u}_n(x_1, t_1) - \underline{u}_n(x_0, t_0)| = |\underline{u}_{nt}(\xi, \eta)(t_1 - t_0) + \underline{u}_{nx}(\xi, \eta)(x_1 - x_0)| \tag{8.31}$$

with $(\xi, \eta) \in D_\delta$ lying on the straight line joining (x_0, t_0) to (x_1, t_1). It follows from (8.31) and Proposition 8.14, that

$$\begin{aligned}
|\underline{u}_n&(x_1, t_1) - \underline{u}_n(x_0, t_0)| \\
&\leq |\underline{u}_{nt}(\xi, \eta)||t_1 - t_0| + |\underline{u}_{nx}(\xi, \eta)||x_1 - x_0| \\
&\leq \max\left\{ \frac{2c'}{\sqrt{\pi}}(1 + \delta^{1/2}) + M_0', \; \frac{2^{\alpha+1}I_\alpha}{\alpha\sqrt{\pi}}k_\delta(1 + \delta^{\alpha/2}) + c' + M_0'' \right\} \\
&\quad \times (|t_1 - t_0| + |x_1 - x_0|) \\
&\leq \max\left\{ \frac{2c'}{\sqrt{\pi}}(1 + \delta^{1/2}) + M_0', \; \frac{2^{\alpha+1}I_\alpha}{\alpha\sqrt{\pi}}k_\delta(1 + \delta^{\alpha/2}) + c' + M_0'' \right\} \\
&\quad \times \sqrt{2}\,(|(x_1, t_1) - (x_0, t_0)|).
\end{aligned} \tag{8.32}$$

Since (8.32) holds for all (x_1, t_1), $(x_0, t_0) \in D_\delta$, and $\underline{u}_n$ is continuous on $\bar{D}_\delta$, then it follows that (8.32) holds for all (x_1, t_1), (x_0, t_0) on $\bar{D}_\delta$. It is then an immediate consequence of (8.32) that the sequence of functions $\{\underline{u}_n\}_{n \in \mathbb{N}}$ are uniformly equicontinuous on $\bar{D}_\delta$. Moreover, it follows from (8.30) that $\{\underline{u}_n\}_{n \in \mathbb{N}}$ are uniformly bounded (by m_0) on $\bar{D}_\delta$. It then follows immediately from the *Ascoli–Arzéla compactness criterion* (see, for example, [66][p. 154–158]) that there exists a subsequence $\{\underline{u}_{n_j}\}_{j \in \mathbb{N}}$ ($1 \leq n_1 < n_2 < n_3 < \cdots$ and $n_j \to \infty$ as $j \to \infty$) and a continuous function $u_c : \bar{D}_\delta \to \mathbb{R}$ such that

$$\underline{u}_{n_j} \to u_c \text{ as } j \to \infty \text{ uniformly on } \bar{D}_\delta^{0,X}, \tag{8.33}$$

for any $X > 0$. From (8.33), we have that for each $(x, t) \in \bar{D}_\delta$, the real sequence $\{\underline{u}_{n_j}(x, t)\}_{n_j \in \mathbb{N}}$, is such that

$$\underline{u}_{n_j}(x, t) \to u_c(x, t) \text{ as } j \to \infty. \tag{8.34}$$

It also follows from (8.28) (convergence of subsequences of convergent real sequences) that

$$\underline{u}_{n_j}(x, t) \to u^*(x, t) \text{ as } j \to \infty. \tag{8.35}$$

It follows from (8.34) and (8.35) (uniqueness of limits of convergent real sequences) that

$$u^*(x, t) = u_c(x, t); \quad \forall \, (x, t) \in \bar{D}_\delta$$

and so $u^* : \bar{D}_\delta \to \mathbb{R}$ is continuous and via (8.33),

$$\underline{u}_{n_j} \to u^* \text{ as } j \to \infty \text{ uniformly on } \bar{D}_\delta^{0,X},$$

for any $X > 0$, as required. $\square$

As a consequence we have,

Corollary$_\ddagger$ 8.20 *For any $X > 0$,*

$$\underline{u}_n \to u^* \text{ as } n \to \infty \text{ uniformly on } \bar{D}_\delta^{0,X}.$$

Proof From Lemma 8.19, we have

$$\underline{u}_{n_j} \to u^* \text{ as } j \to \infty \text{ uniformly on } \bar{D}_\delta^{0,X}, \tag{8.36}$$

for any $X > 0$. Thus, given any $\epsilon > 0$, there exists $J_\epsilon \in \mathbb{N}$ (independent of $(x, t) \in \bar{D}_\delta^{0,X}$) such that for all $j \geq J_\epsilon$,

$$|\underline{u}_{n_j}(x, t) - u^*(x, t)| < \epsilon; \quad \forall \, (x, t) \in \bar{D}_\delta^{0,X}. \tag{8.37}$$

It now follows from Proposition 8.16 and (8.28) that for any $n \geq n_{(J_\epsilon + 1)}$,

$$0 \leq u^*(x, t) - \underline{u}_n(x, t) \leq u^*(x, t) - \underline{u}_{n_{J_\epsilon}}(x, t); \quad \forall \, (x, t) \in \bar{D}_\delta^{0,X}. \tag{8.38}$$

Thus, via (8.37) and (8.38), we have that for all $n \geq n_{(J_\epsilon + 1)}$, then

$$|\underline{u}_n(x, t) - u^*(x, t)| \leq |\underline{u}_{n_{J_\epsilon}}(x, t) - u^*(x, t)| < \epsilon; \quad \forall \, (x, t) \in \bar{D}_\delta^{0,X}.$$

Thus, it follows that

$$\underline{u}_n \to u^* \text{ as } n \to \infty \text{ uniformly on } \bar{D}_\delta^{0,X},$$

as required. $\square$

Proposition$_\ddagger$ 8.21 *Let $u : \bar{D}_\delta \to \mathbb{R}$ be any solution to (B-R-D-C). Then,*

$$u^*(x, t) \leq u(x, t); \quad \forall (x, t) \in \bar{D}_\delta.$$

Proof It follows from Proposition 8.17 that for each $n \in \mathbb{N}$,

$$\underline{u}_n(x, t) \leq u(x, t); \quad \forall (x, t) \in \bar{D}_\delta.$$

It then follows from (8.28) that

$$u^*(x, t) \leq u(x, t); \quad \forall (x, t) \in \bar{D}_\delta,$$

as required. $\square$

Remark 8.22 $u^* : \bar{D}_\delta \to \mathbb{R}$ is continuous and from (8.29),

$$|u^*(x,t)| \leq m_0,$$

for all $(x,t) \in \bar{D}_\delta$, so u^* is bounded on $\bar{D}_\delta$. It follows that

$$u^* \in B_A^\delta \tag{8.39}$$

and

$$\|u^*\|_A \leq m_0. \qquad \lrcorner$$

With Remark 8.22 it remains to establish that $u^* : \bar{D}_\delta \to \mathbb{R}$ satisfies the appropriate integral equation in the Hölder Equivalence Lemma 5.10. To begin, we introduce the functions $\underline{v}_n : \bar{D}_\delta \to \mathbb{R}$ ($n \in \mathbb{N}$) and $v : \bar{D}_\delta \to \mathbb{R}$, as follows,

$$\underline{v}_n(x,t) = \frac{1}{\sqrt{\pi}} \int_{-\infty}^{\infty} \left(u_0 \left(x + 2\sqrt{t}\lambda \right) - \frac{1}{2n} \right) e^{-\lambda^2} d\lambda,$$

$$v(x,t) = \frac{1}{\sqrt{\pi}} \int_{-\infty}^{\infty} u_0 \left(x + 2\sqrt{t}\lambda \right) e^{-\lambda^2} d\lambda \tag{8.40}$$

for all $(x,t) \in \bar{D}_\delta$. We note that v_n and v are well-defined and $\underline{v}_n, v \in B_A^\delta$ (Lemma 5.8). Moreover,

$$\lim_{n \to \infty} \underline{v}_n(x,t) = \lim_{n \to \infty} \left(v(x,t) - \frac{1}{\sqrt{\pi}} \int_{-\infty}^{\infty} \frac{1}{2n} e^{-\lambda^2} d\lambda \right) = v(x,t); \quad \forall (x,t) \in \bar{D}_\delta. \tag{8.41}$$

It remains only to consider the functions $\underline{w}_n : \bar{D}_\delta \to \mathbb{R}$ ($n \in \mathbb{N}$) and $w : \bar{D}_\delta \to \mathbb{R}$ defined as follows,

$$\underline{w}_n(x,t) = \frac{1}{\sqrt{\pi}} \int_0^t \int_{-\infty}^{\infty} \underline{f}_n \left(\underline{u}_n \left(x + 2\sqrt{t-\tau}\lambda, \tau \right) \right) e^{-\lambda^2} d\lambda d\tau,$$

$$w(x,t) = \frac{1}{\sqrt{\pi}} \int_0^t \int_{-\infty}^{\infty} f \left(u^* \left(x + 2\sqrt{t-\tau}\lambda, \tau \right) \right) e^{-\lambda^2} d\lambda d\tau \tag{8.42}$$

for all $(x,t) \in \bar{D}_\delta$. We note that these functions are well-defined, via Lemma 5.8, since $\underline{u}_n \in B_A^\delta$ ($n \in \mathbb{N}$) (as it is a solution to (B-R-D-C)$_n^l$) and $u^* \in B_A^\delta$ (via Remark 8.22). Moreover $w, \underline{w}_n \in B_A^\delta$ ($n \in \mathbb{N}$), via Lemma 5.8. We also observe that, $\underline{f}_n(\underline{u}_n)$, $f(u^*) \in B_A^\delta$, and

$$\|\underline{f}_n(\underline{u}_n)\|_A \leq c', \quad \|f(u^*)\|_A \leq c' \tag{8.43}$$

for all $n \in \mathbb{N}$, via Remark 8.22. We now have,

Lemma‡ 8.23 *For each* $(x, t) \in \bar{D}_\delta$, *the real sequence* $\{\underline{w}_n(x, t)\}_{n \in \mathbb{N}}$ *is convergent, and*

$$\lim_{n \to \infty} \underline{w}_n(x, t) = w(x, t).$$

Proof Given any $\epsilon > 0$, take

$$\lambda_\epsilon = \max \left\{ \frac{8c'(1 + \delta)}{\sqrt{\pi}\epsilon}, 1 \right\}. \tag{8.44}$$

Now fix $(x, t) \in \bar{D}_\delta$, then

$$
\begin{aligned}
&|\underline{w}_n(x, t) - w(x, t)| \\
&\leq \frac{1}{\sqrt{\pi}} \int_0^t \int_{-\infty}^{\infty} |\underline{f}_n \left(\underline{u}_n \left(x + 2\sqrt{t - \tau}\lambda, \tau\right)\right) \\
&\qquad - f \left(u^* \left(x + 2\sqrt{t - \tau}\lambda, \tau\right)\right)| e^{-\lambda^2} d\lambda d\tau \\
&\leq \frac{1}{\sqrt{\pi}} \int_0^t \int_{-\lambda_\epsilon}^{\lambda_\epsilon} |\underline{f}_n \left(\underline{u}_n \left(x + 2\sqrt{t - \tau}\lambda, \tau\right)\right) \\
&\qquad - f \left(u^* \left(x + 2\sqrt{t - \tau}\lambda, \tau\right)\right)| e^{-\lambda^2} d\lambda d\tau \\
&\quad + \frac{1}{\sqrt{\pi}} \int_0^t \int_{\lambda_\epsilon}^{\infty} |\underline{f}_n \left(\underline{u}_n \left(x + 2\sqrt{t - \tau}\lambda, \tau\right)\right)| e^{-\lambda^2} d\lambda d\tau \\
&\quad + \frac{1}{\sqrt{\pi}} \int_0^t \int_{\lambda_\epsilon}^{\infty} |f \left(u^* \left(x + 2\sqrt{t - \tau}\lambda, \tau\right)\right)| e^{-\lambda^2} d\lambda d\tau \\
&\quad + \frac{1}{\sqrt{\pi}} \int_0^t \int_{-\infty}^{-\lambda_\epsilon} |\underline{f}_n \left(\underline{u}_n \left(x + 2\sqrt{t - \tau}\lambda, \tau\right)\right)| e^{-\lambda^2} d\lambda d\tau \\
&\quad + \frac{1}{\sqrt{\pi}} \int_0^t \int_{-\infty}^{-\lambda_\epsilon} |f \left(u^* \left(x + 2\sqrt{t - \tau}\lambda, \tau\right)\right)| e^{-\lambda^2} d\lambda d\tau
\end{aligned}
$$

and so

$$
\begin{aligned}
&|\underline{w}_n(x, t) - w(x, t)| \\
&< \frac{1}{\sqrt{\pi}} \int_0^t \int_{-\lambda_\epsilon}^{\lambda_\epsilon} |\underline{f}_n \left(\underline{u}_n \left(x + 2\sqrt{t - \tau}\lambda, \tau\right)\right) \\
&\qquad - f \left(u^* \left(x + 2\sqrt{t - \tau}\lambda, \tau\right)\right)| e^{-\lambda^2} d\lambda d\tau + \frac{\epsilon}{2},
\end{aligned} \tag{8.45}
$$

on using (8.43) and (8.44). Now, via Corollary 8.20, Proposition 8.9 and Proposition 8.11, it follows that there exists $N_\epsilon \in \mathbb{N}$, independent of $(\lambda, \tau) \in [-\lambda_\epsilon, \lambda_\epsilon] \times [0, t]$ such that for all $n \geq N_\epsilon$, then

$$\left| \underline{f}_n \left(u^* \left(x + 2\sqrt{t - \tau}\lambda, \tau \right) \right) - f \left(u^* \left(x + 2\sqrt{t - \tau}\lambda, \tau \right) \right) \right| < \frac{\epsilon}{4\delta},$$

$$\left| \underline{u}_n \left(x + 2\sqrt{t - \tau}\lambda, \tau \right) - u^* \left(x + 2\sqrt{t - \tau}\lambda, \tau \right) \right| \leq \left(\frac{\epsilon}{12 k_H \delta} \right)^{1/\alpha}$$

for all $(\lambda, \tau) \in [-\lambda_\epsilon, \lambda_\epsilon] \times [0, t]$ with $k_H > 0$ being a Hölder constant for $f \in H_\alpha$ on $[-m_0, m_0]$. It then follows from (8.45) that, for all $n \geq N_\epsilon$ (which may depend on $(x, t) \in \bar{D}_\delta$), then via Proposition 8.9,

$$\left| \underline{w}_n(x, t) - w(x, t) \right|$$

$$< \frac{1}{\sqrt{\pi}} \int_0^\delta \int_{-\lambda_\epsilon}^{\lambda_\epsilon} \left(\left| \underline{f}_n \left(\underline{u}_n \left(x + 2\sqrt{t - \tau}\lambda, \tau \right) \right) - \underline{f}_n \left(u^* \left(x + 2\sqrt{t - \tau}\lambda, \tau \right) \right) \right| \right.$$

$$\left. + \left| \underline{f}_n \left(u^* \left(x + 2\sqrt{t - \tau}\lambda, \tau \right) \right) - f \left(u^* \left(x + 2\sqrt{t - \tau}\lambda, \tau \right) \right) \right| \right) e^{-\lambda^2} d\lambda d\tau + \frac{\epsilon}{2}$$

$$\leq \frac{1}{\sqrt{\pi}} \int_0^\delta \int_{-\lambda_\epsilon}^{\lambda_\epsilon} \left(3 k_H |\underline{u}_n - u^*|^\alpha \left(x + 2\sqrt{t - \tau}\lambda, \tau \right) + \frac{\epsilon}{4\delta} \right) e^{-\lambda^2} d\lambda d\tau + \frac{\epsilon}{2}$$

$$\leq \frac{1}{\sqrt{\pi}} \int_0^\delta \int_{-\lambda_\epsilon}^{\lambda_\epsilon} \left(\frac{\epsilon}{4\delta} + \frac{\epsilon}{4\delta} \right) e^{-\lambda^2} d\lambda d\tau + \frac{\epsilon}{2}$$

$$\leq \frac{\epsilon}{2\delta\sqrt{\pi}} \int_0^\delta \int_{-\infty}^{\infty} e^{-\lambda^2} d\lambda d\tau + \frac{\epsilon}{2}$$

$$\leq \frac{\epsilon}{2} + \frac{\epsilon}{2}$$

$$\leq \epsilon.$$

Therefore for each $(x, t) \in \bar{D}_\delta$, the real sequence $\{\underline{w}_n(x, t)\}_{n \in \mathbb{N}}$ is convergent and

$$\lim_{n \to \infty} \underline{w}_n(x, t) = w(x, t),$$

as required. $\square$

We now have,

Lemma$_\ddagger$ 8.24 *The function $u^* : \bar{D}_\delta \to \mathbb{R}$ is such that, $u^* \in B_A^\delta$, and*

$$u^*(x, t) = \frac{1}{\sqrt{\pi}} \int_{-\infty}^{\infty} u_0 \left(x + 2\sqrt{t}\lambda \right) e^{-\lambda^2} d\lambda$$

$$+ \frac{1}{\sqrt{\pi}} \int_0^t \int_{-\infty}^{\infty} f \left(u^* \left(x + 2\sqrt{t - \tau}\lambda, \tau \right) \right) e^{-\lambda^2} d\lambda d\tau$$

for all $(x, t) \in \bar{D}_\delta$.

Proof For each $n \in \mathbb{N}$, then by construction $\underline{u}_n : \bar{D}_\delta \to \mathbb{R}$ is a solution to (B-R-D-C)_n^l on $\bar{D}_\delta$. Since, for each $n \in \mathbb{N}$, (B-R-D-C)_n^l has $\underline{f}_n \in H_\alpha$, it follows from the Hölder Equivalence Lemma 5.10 that $\underline{u}_n \in B_A^\delta$ and

$$
\begin{aligned}
\underline{u}_n(x, t) &= \frac{1}{\sqrt{\pi}} \int_{-\infty}^{\infty} \left(u_0 \left(x + 2\sqrt{t}\lambda \right) - \frac{1}{2n} \right) e^{-\lambda^2} d\lambda \\
&\quad + \frac{1}{\sqrt{\pi}} \int_0^t \int_{-\infty}^{\infty} \underline{f}_n \left(\underline{u}_n \left(x + 2\sqrt{t - \tau}\lambda, \tau \right) \right) e^{-\lambda^2} d\lambda d\tau \\
&= \underline{v}_n(x, t) + \underline{w}_n(x, t)
\end{aligned}
\tag{8.46}
$$

for all $(x, t) \in \bar{D}_\delta$. Now fix $(x, t) \in \bar{D}_\delta$. It then follows from (8.46), (8.28), (8.41) and Lemma 8.23 that

$$
u^*(x, t) = v(x, t) + w(x, t); \quad \forall (x, t) \in \bar{D}_\delta,
$$

which, via (8.40) and (8.42) becomes

$$
\begin{aligned}
u^*(x, t) &= \frac{1}{\sqrt{\pi}} \int_{-\infty}^{\infty} u_0 \left(x + 2\sqrt{t}\lambda \right) e^{-\lambda^2} d\lambda \\
&\quad + \frac{1}{\sqrt{\pi}} \int_0^t \int_{-\infty}^{\infty} f \left(u^* \left(x + 2\sqrt{t - \tau}\lambda, \tau \right) \right) e^{-\lambda^2} d\lambda d\tau
\end{aligned}
$$

for all $(x, t) \in \bar{D}_\delta$. In addition, via Remark 8.22, $u^* \in B_A^\delta$. The proof is complete. $\qquad \square$

It now follows immediately from Lemma 8.24 and the Hölder Equivalence Lemma 5.10 that $u^* : \bar{D}_\delta \to \mathbb{R}$ provides a solution to (B-R-D-C) on $\bar{D}_\delta$. That $u^* : \bar{D}_\delta \to \mathbb{R}$ is a minimal solution to (B-R-D-C) on $\bar{D}_\delta$ follows from Proposition 8.21 and the bound follows from Remark 8.22. The proof of Theorem 8.3 is complete.

A global existence theorem can now be established.

Theorem$_\ddagger$ 8.25 *Consider (B-R-D-C) with $f \in H_\alpha$ for some $\alpha \in (0, 1)$. When (B-R-D-C) is a priori bounded on $\bar{D}_T$ for any $0 \leq T \leq T'$, then (B-R-D-C) has a constructed minimal and a constructed maximal solution on $\bar{D}_{T'}$.*

Proof The proof is a direct application of Theorem 8.3, with the a priori bounds allowing $[0, T']$ to be covered in a finite number of steps. The details are as in the proof of Theorem 6.4, using Remark 8.4. $\qquad \square$

Following Proposition 8.17 we also have the following comparison-type result.

Proposition$_{\ddagger}$ 8.26 *Let $f \in H_\alpha$ for some $\alpha \in (0, 1)$, and let $\underline{w}, \overline{w} : \bar{D}_T \to \mathbb{R}$ be a regular subsolution and a regular supersolution to (B-R-D-C), respectively. Let $\underline{u}^c, \overline{u}^c : \bar{D}_T \to \mathbb{R}$ be constructed minimal and maximal solutions to (B-R-D-C), then*

$$\underline{u}^c(x, t) \leq \overline{w}(x, t) \quad and \quad \overline{u}^c(x, t) \geq \underline{w}(x, t)$$

for all $(x, t) \in \bar{D}_T$.

Proof We give a proof for the first inequality. The second inequality follows the same argument, with obvious modifications. Now, $\underline{u}^c : \bar{D}_T \to \mathbb{R}$ is a constructed minimal solution to (B-R-D-C). It follows, via Remark 8.4, and the construction of $\underline{u}^c$, that Proposition 8.17 holds on each constructional subdomain of $\bar{D}_T$ in turn. The result then follows. $\square$

Remark 8.27 We observe that when uniqueness holds for (B-R-D-C) on $\bar{D}_T$, then $\underline{u}^c = \overline{u}^c$ on $\bar{D}_T$ and Proposition 8.26 becomes a full Comparison Theorem for (B-R-D-C). $\lrcorner$

The issue we have not addressed this far is uniqueness, and we may anticipate that general uniqueness, where $f \in H_\alpha$, for $\alpha \in (0, 1)$, is false, via the following example.

Example 8.28 Consider the (B-R-D-C) problem where $f : \mathbb{R} \to \mathbb{R}$ is such that $f = [u^p]^+$ for some $p \in (0, 1)$ and $u_0 : \mathbb{R} \to \mathbb{R}$ is such that $u_0(x) = 0$ for all $x \in \mathbb{R}$. Simple calculations show that $f \in H_p \backslash L_u$ and $u_0 \in \text{BPC}^2(\mathbb{R})$. Now define $u_1, u_2 : \bar{D}_T \to \mathbb{R}$ for any $T > 0$ to be

$$u_1(x, t) = 0; \quad \forall (x, t) \in \bar{D}_T,$$

$$u_2(x, t) = \begin{cases} 0 & ; (x, t) \in \mathbb{R} \times [0, t_s] \\ ((1 - p)(t - t_s))^{1/(1-p)} & ; \mathbb{R} \times (t_s, T], \end{cases}$$

for any $0 \leq t_s < T$. It is readily verified that u_1 and u_2 are distinct solutions to (B-R-D-C). $\lrcorner$

We next consider a further pathological example to illustrate the breadth of Theorem 8.3, where the reaction function is non-Lipschitz on every closed bounded interval.

Example‡ 8.29 Consider (B-R-D-C) with reaction function $f_{\alpha,b} : \mathbb{R} \to \mathbb{R}$ given by

$$f_{\alpha,b}(u) = \sum_{n=0}^{\infty} b^{-n\alpha} \cos(b^n u) \tag{8.47}$$

for all $u \in \mathbb{R}$, where $b > 1$ and $\alpha \in (0, 1)$. This function was used by Weierstrass [73], to exhibit the existence of a real valued function which is everywhere continuous, but non-differentiable almost everywhere. As a consequence of Rademacher's Theorem [38] (p.100), this function is not Lipschitz continuous on any closed bounded interval. However, for any $\alpha' \in (0, \alpha)$,

$$|f_{\alpha,b}(u) - f_{\alpha,b}(v)| \leq \sum_{n=0}^{\infty} b^{-n\alpha} \left| \cos(b^n u) - \cos(b^n v) \right|$$

$$\leq 2 \sum_{n=0}^{\infty} b^{-n\alpha} \left| b^n u - b^n v \right|^{\alpha'}$$

$$= 2 \sum_{n=0}^{\infty} b^{n(\alpha'-\alpha)} |u - v|^{\alpha'}$$

$$= \frac{2}{(1 - b^{(\alpha'-\alpha)})} |u - v|^{\alpha'}$$

for any $u, v \in \mathbb{R}$. Hence $f_{\alpha,b} \in H_{\alpha'}$ with Hölder constant $\dfrac{2}{(1 - b^{(\alpha'-\alpha)})}$ on any closed bounded interval. Moreover, f is bounded on $\mathbb{R}$ with

$$|f_{\alpha,b}(u)| \leq \frac{1}{(1 - b^{-\alpha})}; \quad \forall u \in \mathbb{R}. \tag{8.48}$$

Now let $u : \bar{D}_T \to \mathbb{R}$ be any solution to (B-R-D-C), and let $w_+ : \bar{D}_T \to \mathbb{R}$ and $w_- : \bar{D}_T \to \mathbb{R}$ be such that

$$w_+(x, t) = \frac{t}{(1 - b^{-\alpha})} + \sup_{\lambda \in \mathbb{R}} u_0(\lambda)$$

$$w_-(x, t) = \frac{-t}{(1 - b^{-\alpha})} + \inf_{\lambda \in \mathbb{R}} u_0(\lambda)$$

for all $(x, t) \in \bar{D}_T$. Then,

$$u_t - u_{xx} - \frac{1}{(1 - b^{-\alpha})} = f_{\alpha,b}(u) - \frac{1}{(1 - b^{-\alpha})} \leq 0$$

$$w_{+t} - w_{+xx} - \frac{1}{(1 - b^{-\alpha})} = 0 \geq 0$$

for all $(x, t) \in D_T$. It follows via Comparison Theorem 7.1, that

$$u(x, t) \leq w_+(x, t); \quad \forall(x, t) \in \bar{D}_T.$$

Similarly, we establish that

$$u(x, t) \geq w_-(x, t); \quad \forall(x, t) \in \bar{D}_T.$$

Thus,

$$\frac{-T}{(1 - b^{-\alpha})} - \|u_0\|_B \leq u(x, t) \leq \frac{T}{(1 - b^{-\alpha})} + \|u_0\|_B; \quad \forall(x, t) \in \bar{D}_T,$$

and so

$$\|u\|_A \leq \frac{T}{(1 - b^{-\alpha})} + \|u_0\|_B.$$

We conclude that (B-R-D-C) is a priori bounded on $\bar{D}_T$ for any $T > 0$. Thus (B-R-D-C) has a global constructed minimal solution $\underline{u}^c : \bar{D}_\infty \to \mathbb{R}$ and a global constructed maximal solution $\bar{u}^c : \bar{D}_\infty \to \mathbb{R}$, via Theorem 8.25, and

$$\frac{-t}{(1 - b^{-\alpha})} + \inf_{\lambda \in \mathbb{R}} u_0(\lambda) \leq \underline{u}^c(x, t) \leq \bar{u}^c(x, t) \leq \frac{t}{(1 - b^{-\alpha})} + \sup_{\lambda \in \mathbb{R}} u_0(\lambda). \quad \lrcorner$$

Additionally, we have,

Example 8.30 Consider the (B-R-D-C) with reaction function $f \in H_{\min\{p,q\}} \cap L_u$, given by

$$f(u) = [u^p]^+ (u - 1/2) [(1 - u)^q]^+; \quad \forall u \in \mathbb{R},$$

with $p, q \in (0, 1)$. We can now elaborate on the conclusions made in Example 7.12.

(i) (B-R-D-C) is a priori bounded on $\bar{D}_T$ for any $T \geq 0$, via a simple application of Theorem 7.1.

(ii) There exists a unique global solution for all $u_0 \in \mathrm{BPC}^2(\mathbb{R})$ via (i) and Theorem 8.25, and Theorem 7.2.

(iii) (B-R-D-C) is globally well-posed on $\mathrm{BPC}^2(\mathbb{R})$ via (ii) and Corollary 7.8.

(iv) Let I be a closed bounded interval such that $I \subset (-\infty, 1/2)$ or $I \subset (1/2, \infty)$. Then (B-R-D-C) is uniformly globally well-posed on $A_I(\mathbb{R})$ via (ii) and Corollary 7.11. $\lrcorner$

In conclusion we remark that the approach adopted here in the proof of Theorem 8.3 was primarily motivated by the specific problem in [54] and Chapter 9 of this monograph. However, the methodology is remarkably similar to that developed in the context of ordinary differential equations in

Carathéodory [16]. Carathéodory's approach has been used in [17] (p.45) to establish an analogous result to Theorem 8.3 for the ordinary differential equation problem

$$u_t = f(u, t), \quad u(0) = u_0$$

on $t \in [0, T]$ with f a continuous function in both variables. The methodology is similar in the sense that successive approximations are made and the Ascoli–Arzela compactness theorem is used to establish the existence of a limit. In addition, global existence results for second order parabolic partial differential equations (similar to Theorem 8.25), under various hypotheses, are available in [61] and [14]. The results in [14] are obtained by examining the limit of a sequence of Dirichlet problems with expanding domains together with the theory developed in [69] and [53] to guarantee existence and regularity of solutions to the approximating Dirichlet problems (under the assumption of the existence of global supersolutions and subsolutions).

We complete this section by establishing structural qualitative features of maximal and minimal solutions to (B-R-D-C). The first three results do not require that the associated maximal and minimal solution be constructed, whereas the fourth and fifth results require that the associated maximal and minimal solution is constructed. Once qualitative features have been established, we provide a conditional continuous dependence result for solutions to (B-R-D-C). To begin, we have,

Proposition 8.31 *Consider (B-R-D-C) with $f \in H_\alpha$ for some $\alpha \in (0, 1)$, and $u_0 \in BPC^2(\mathbb{R})$ such that $u_0 : \mathbb{R} \to \mathbb{R}$ is constant. Suppose that $\overline{u}, \underline{u} : \bar{D}_T \to \mathbb{R}$ are a maximal solution and a minimal solution to (B-R-D-C) on $\bar{D}_T$, respectively. Then $\overline{u}(x, t)$ and $\underline{u}(x, t)$ are independent of $x \in \mathbb{R}$ for each $t \in [0, T]$.*

Proof First, for each $k \in \mathbb{R}$, consider $w^{(k)} : \bar{D}_T \to \mathbb{R}$ given by,

$$w^{(k)}(x, t) = \overline{u}(x + k, t); \quad \forall (x, t) \in \bar{D}_T. \tag{8.49}$$

It follows immediately, that for each $k \in \mathbb{R}$, $w^{(k)}$ is a solution to (B-R-D-C) on $\bar{D}_T$. Now, since $\overline{u} : \bar{D}_T \to \mathbb{R}$ is a maximal solution to (B-R-D-C) on $\bar{D}_T$, it follows from Definition 8.1 that

$$w^{(k)}(x, t) \leq \overline{u}(x, t); \quad \forall (x, t) \in \bar{D}_T, \tag{8.50}$$

and so,

$$\overline{u}(x + k, t) \leq \overline{u}(x, t); \quad \forall (x, t, k) \in \bar{D}_T \times \mathbb{R}. \tag{8.51}$$

Now, take $(x_1, t), (x_2, t) \in \bar{D}_T$, and set $x = x_2$ and $k = x_1 - x_2$ in (8.51), to obtain

$$\bar{u}(x_1, t) \leq \bar{u}(x_2, t). \tag{8.52}$$

Next set $x = x_1$ and $k = x_2 - x_1$ in (8.51) to obtain

$$\bar{u}(x_2, t) \leq \bar{u}(x_1, t). \tag{8.53}$$

It follows from (8.52) and (8.53) that $\bar{u}(x_1, t) = \bar{u}(x_2, t)$, as required. The result for $\underline{u} : \bar{D}_T \to \mathbb{R}$ follows by a symmetrical argument. $\qquad \square$

Remark 8.32 It follows from Proposition 8.31 that for (B-R-D-C) with $f \in H_\alpha$ for some $\alpha \in (0, 1)$, and $u_0 \in \mathrm{BPC}^2(\mathbb{R})$ such that u_0 is constant ($u_0 = c$), then a maximal solution $\bar{u} : \bar{D}_T \to \mathbb{R}$ and a minimal solution $\underline{u} : \bar{D}_T \to \mathbb{R}$ to (B-R-D-C) on $\bar{D}_T$ will be given by a maximal solution $\bar{U} : [0, T] \to \mathbb{R}$ and a minimal solution $\underline{U} : [0, T] \to \mathbb{R}$ to the following initial value problem,

$$u_t = f(u), \quad \forall t \in (0, T], \quad u(0) = c.$$

The existence of $\bar{U}, \underline{U} : [0, T] \to \mathbb{R}$ are guaranteed by Proposition 8.31. $\qquad \lrcorner$

We also have,

Proposition 8.33 *Consider (B-R-D-C) with $f \in H_\alpha$ for some $\alpha \in (0, 1)$ and $u_0 \in \mathrm{BPC}^2(\mathbb{R})$ such that*

$$u_0(x) = u_0(-x); \quad \forall x \in \mathbb{R}.$$

Suppose that $\bar{u}, \underline{u} : \bar{D}_T \to \mathbb{R}$ are a maximal solution and a minimal solution to (B-R-D-C) on $\bar{D}_T$, respectively. Then,

$$\bar{u}(x, t) = \bar{u}(-x, t) \ \text{and} \ \underline{u}(x, t) = \underline{u}(-x, t); \quad \forall (x, t) \in \bar{D}_T.$$

Proof First introduce $w : \bar{D}_T \to \mathbb{R}$ as

$$w(x, t) = \bar{u}(-x, t); \quad \forall (x, t) \in \bar{D}_T. \tag{8.54}$$

Then, it follows that

$$w(x, 0) = u_0(x); \quad \forall x \in \mathbb{R}. \tag{8.55}$$

Additionally, it follows from (8.54) that

$$w_t - w_{xx} - f(w) = 0; \quad \forall (x, t) \in D_T. \tag{8.56}$$

Therefore, via (8.55) and (8.56), $w : \bar{D}_T \rightarrow \mathbb{R}$ is a solution to (B-R-D-C) on $\bar{D}_T$. Thus, via (8.54) and Definition 8.1, we have

$$\bar{u}(-x, t) = w(x, t) \leq \bar{u}(x, t); \quad \forall(x, t) \in \bar{D}_T. \tag{8.57}$$

Upon considering $(x, t) = (\pm s, t)$ in (8.57), it follows that

$$\bar{u}(s, t) = \bar{u}(-s, t); \quad \forall(s, t) \in \bar{D}_T,$$

as required. The result for $\underline{u}$ follows a symmetrical argument. $\square$

Additionally, we have,

Proposition 8.34 *Consider (B-R-D-C) with $f \in H_\alpha$ for some $\alpha \in (0, 1)$ and $u_0 \in BPC^2(\mathbb{R})$ such that for some $X > 0$,*

$$u_0(x) = u_0(x + X); \quad \forall x \in \mathbb{R}.$$

Suppose that $\bar{u}, \underline{u} : \bar{D}_T \rightarrow \mathbb{R}$ are a maximal solution and a minimal solution to (B-R-D-C) on $\bar{D}_T$, respectively. Then,

$$\bar{u}(x, t) = \bar{u}(x + X, t) \text{ and } \underline{u}(x, t) = \underline{u}(x + X, t); \quad \forall(x, t) \in \bar{D}_T.$$

Proof First, consider $w : \bar{D}_T \rightarrow \mathbb{R}$ given by

$$w(x, t) = \bar{u}(x + X, t); \quad \forall(x, t) \in \bar{D}_T. \tag{8.58}$$

It follows immediately, that $w : \bar{D}_T \rightarrow \mathbb{R}$ is a solution to (B-R-D-C) on $\bar{D}_T$. Now, since $\bar{u}$ is a maximal solution to (B-R-D-C) on $\bar{D}_T$, it follows from Definition 8.1 that,

$$w(x, t) \leq \bar{u}(x, t); \quad \forall(x, t) \in \bar{D}_T,$$

and so,

$$\bar{u}(x + X, t) \leq \bar{u}(x, t); \quad \forall(x, t) \in \bar{D}_T. \tag{8.59}$$

Similarly, we may establish that,

$$\bar{u}(x - X, t) \leq \bar{u}(x, t); \quad \forall(x, t) \in \bar{D}_T. \tag{8.60}$$

Now, put $(x, t) = (s, t)$ in (8.59) and $(x, t) = (s + X, t)$ in (8.60), from which it follows that,

$$\bar{u}(x + X, t) = \bar{u}(x, t); \quad \forall(x, t) \in \bar{D}_T,$$

as required. The result for $\underline{u}$ follows a symmetrical argument. $\square$

Remark 8.35 We note here, that if we consider (B-R-D-C) with $f \in H_\alpha$, for some $\alpha \in (0, 1)$, and initial data $u_0 \in \mathrm{BPC}^2(\mathbb{R})$ such that

$$u_0(x) = -u_0(-x); \quad \forall x \in \mathbb{R},$$

then the method of proof adopted in the above propositions fails to establish a similar conclusion for a maximal or a minimal solution to (B-R-D-C) on $\bar{D}_T$. ⌐

We now consider a result concerning constructed maximal and constructed minimal solutions.

Proposition 8.36 *Consider (B-R-D-C) with $f \in H_\alpha$ for some $\alpha \in (0, 1)$, and $u_0 \in \mathrm{BPC}^2(\mathbb{R})$ such that u_0 is non-decreasing (non-increasing). Suppose that $\bar{u}^c, \underline{u}^c : \bar{D}_T \to \mathbb{R}$ are a constructed maximal solution and a constructed minimal solution to (B-R-D-C) on $\bar{D}_T$, respectively. Then $\bar{u}^c(x, t)$ and $\underline{u}^c(x, t)$ are non-decreasing (non-increasing) with $x \in \mathbb{R}$ for each $t \in [0, T]$.*

Proof We give a proof in the non-decreasing case, with the non-increasing case following similarly. First, for each $k \geq 0$, introduce $\underline{w}^{(k)} : \bar{D}_T \to \mathbb{R}$ such that

$$\underline{w}^{(k)}(x, t) = \bar{u}^c(x - k, t); \quad \forall (x, t, k) \in \bar{D}_T \times \bar{\mathbb{R}}^+. \qquad (8.61)$$

It follows that $\underline{w}^{(k)} : \bar{D}_T \to \mathbb{R}$ is a regular subsolution to (B-R-D-C) on $\bar{D}_T$. An application of Proposition 8.26 then gives,

$$\bar{u}^c(x - k, t) = \underline{w}^{(k)}(x, t) \leq \bar{u}^c(x, t); \quad \forall (x, t, k) \in \bar{D}_T \times \bar{\mathbb{R}}^+. \qquad (8.62)$$

Now take $(x_1, t), (x_2, t) \in \bar{D}_T$, with $x_2 \geq x_1$, and set $x = x_2$ with $k = x_2 - x_1$ in (8.62), to obtain

$$\bar{u}^c(x_1, t) \leq \bar{u}^c(x_2, t); \quad \forall x_2 \geq x_1, \ t \in [0, T].$$

Therefore, $\bar{u}^c(x, t)$ is non-decreasing with $x \in \mathbb{R}$ for each $t \in [0, T]$. The result for $\underline{u}^c$ follows a symmetrical argument. □

We now consider the behavior of solutions to (B-R-D-C) as $|x| \to \infty$.

Proposition 8.37 *Consider (B-R-D-C) with $f \in H_\alpha$ for some $\alpha \in (0, 1)$, and $u_0 \in \mathrm{BPC}^2(\mathbb{R})$ such that*

$$\lim_{x \to \pm\infty} u_0(x) = u_0^{\pm}.$$

Let $\overline{u}^c, \underline{u}^c : \bar{D}_\delta \to \mathbb{R}$ be the constructed maximal solution and the constructed minimal solution to (B-R-D-C) on $\bar{D}_\delta$, as given by Theorem 8.3, respectively. Then,

$$\limsup_{x \to \pm\infty} \overline{u}^c(x, t) \leq \overline{U}^\pm(t), \quad \liminf_{x \to \pm\infty} \underline{u}^c(x, t) \geq \underline{U}^\pm(t)$$

uniformly for $t \in [0, \delta]$, where $\overline{U}^\pm : [0, \delta] \to \mathbb{R}$ is the maximal solution and $\underline{U}^\pm : [0, \delta] \to \mathbb{R}$ is the minimal solution respectively, to the initial value problem,

$$U_t^\pm = f(U^\pm); \quad \forall t \in (0, \delta], \quad U^\pm(0) = u_0^\pm. \tag{8.63}$$

Proof First note that the existence of $\overline{U}^\pm, \underline{U}^\pm : [0, \delta] \to \mathbb{R}$ is guaranteed by [17, Theorem 2.3, p.10]. We consider the case $x \to +\infty$, with the case $x \to -\infty$ following similarly. To begin, let $\delta > 0$ and $\underline{u}^c : \bar{D}_\delta \to \mathbb{R}$ be as in Theorem 8.3. Moreover, let $\underline{u}_n : \bar{D}_\delta \to \mathbb{R}$ be the solution to the problem (B-R-D-C))$_n^l$ for each $n \in \mathbb{N}$, as employed in the construction of Theorem 8.3, via Proposition 8.13. It follows immediately from Remark 8.12 and Remark 8.18 that

$$-m_0 \leq \underline{u}_n(x, t) \leq \underline{u}^c(x, t); \quad \forall(x, t) \in \bar{D}_\delta, \tag{8.64}$$

with m_0 as in Theorem 8.3. Moreover, via (8.64),

$$-m_0 \leq \liminf_{x \to \infty} \underline{u}_n(x, t) \leq \liminf_{x \to \infty} \underline{u}^c(x, t) \tag{8.65}$$

uniformly for $t \in [0, \delta]$ and $n \in \mathbb{N}$. Now for $n \in \mathbb{N}$, since $\underline{u}_n : \bar{D}_\delta \to \mathbb{R}$ is bounded and $\underline{f}_n \in L$, it follows from [72, Theorem 5.2, p. 239] that

$$\underline{u}_n(x, t) \to \underline{U}_n^+(t), \tag{8.66}$$

as $x \to +\infty$ uniformly for $t \in [0, \delta]$, where $\underline{U}_n^+ : [0, \delta] \to \mathbb{R}$ is the unique classical solution (see [17, Theorem 2.3, p.10]) to the problem;

$$\underline{U}_{nt}^+ = \underline{f}_n(\underline{U}_n^+); \quad \forall t \in (0, \delta], \quad \underline{U}_n^+(0) = u_0^+ - \frac{1}{2n}. \tag{8.67}$$

Observe that $\underline{U}_n^+ : [0, \delta] \to \mathbb{R}$ is continuous and, via Remark 8.12 and (8.66), is bounded uniformly for $(t, n) \in [0, \delta] \times \mathbb{N}$ with

$$|\underline{U}_n^+(t)| \leq m_0; \quad \forall t \in [0, \delta]. \tag{8.68}$$

Additionally, it follows immediately from Remark 8.10 and (8.67) that

$$|\underline{U}_n^+(t_1) - \underline{U}_n^+(t_2)| \leq \sup_{t \in (0, \delta]} |\underline{U}_{nt}^+(t)||t_1 - t_2| \leq c'|t_1 - t_2|; \quad \forall t_1, t_2 \in [0, \delta],$$

$$\tag{8.69}$$

with c' independent of $n \in \mathbb{N}$, and as in Theorem 8.3. Thus, it follows from (8.68) that the sequence of continuous functions $\{\underline{U}_n^+\}_{n \in \mathbb{N}}$ is uniformly bounded and, via (8.69), uniformly equicontinuous. It then follows immediately from the *Ascoli–Arzéla compactness criterion* ([66](p.154–158)) that there exists a subsequence $\{\underline{U}_{n_j}^+\}_{j \in \mathbb{N}}$ ($1 \leq n_1 < n_2 < n_3 < \cdots$ and $n_j \to \infty$ as $j \to \infty$) and a continuous function $U : [0, \delta] \to \mathbb{R}$ such that

$$\underline{U}_{n_j}^+(t) \to U(t) \text{ as } j \to \infty \text{ uniformly for } t \in [0, \delta], \tag{8.70}$$

and with $U : [0, T] \to \mathbb{R}$ satisfying the bound (8.68) above. Now, let $\epsilon > 0$. Then, via (8.70) and Remark 8.10, there exists $N \in \mathbb{N}$ such that for all $n_j \geq N$,

$$\frac{1}{2n_j} < \frac{\epsilon}{3}, \tag{8.71}$$

$$|\underline{f}_{n_j}(u) - f(u)| < \frac{\epsilon}{3\delta}; \quad \forall u \in [-m_0, m_0], \tag{8.72}$$

$$|\underline{U}_{n_j}^+(\tau) - U(\tau)| < \left(\frac{\epsilon}{3\delta k_H}\right)^{1/\alpha}; \quad \forall \tau \in [0, \delta], \tag{8.73}$$

where k_H is a Hölder constant for $f \in H_\alpha$ on $[-m_0, m_0]$. It now follows from (8.67), (8.71), (8.72) and (8.73) that, with $n_j \geq N$,

$$\left|\underline{U}_{n_j}^+(t) - u_0^+ - \int_0^t f(U(\tau))d\tau\right|$$

$$= \left|\left(u_0^+ - \frac{1}{2n_j} + \int_0^t \underline{f}_{n_j}(\underline{U}_{n_j}^+(\tau))d\tau\right) - u_0^+ - \int_0^t f(U(\tau))d\tau\right|$$

$$\leq \int_0^t |\underline{f}_{n_j}(\underline{U}_{n_j}^+(\tau)) - f(U(\tau))|d\tau + \frac{1}{2n_j}$$

$$< \int_0^t \left(\frac{\epsilon}{3\delta} + \left|f(\underline{U}_{n_j}^+(\tau)) - f(U(\tau))\right|\right)d\tau + \frac{\epsilon}{3}$$

$$\leq \frac{2\epsilon}{3} + \int_0^t k_H|\underline{U}_{n_j}^+(\tau) - U(\tau)|^\alpha d\tau < \frac{2\epsilon}{3} + \frac{\epsilon}{3} = \epsilon$$

for all $t \in [0, \delta]$. Thus, it follows that

$$\underline{U}_{n_j}^+(t) \to u_0^+ + \int_0^t f(U(\tau))d\tau, \tag{8.74}$$

as $j \to \infty$, uniformly for $t \in [0, \delta]$. Therefore, it follows from (8.70), (8.74) and the uniqueness of limits of real sequences, that

$$U(t) = u_0^+ + \int_0^t f(U(\tau))d\tau; \quad \forall t \in [0, \delta], \tag{8.75}$$

and therefore, since U is continuous on $[0, \delta]$, that U is a classical solution of the initial value problem,

$$U'(t) = f(U) \quad \forall t \in (0, \delta], \quad U(0) = u_0^+. \tag{8.76}$$

Now, suppose that $V : [0, \delta] \to \mathbb{R}$ is a solution to (8.76), and set $M_v = \sup_{t \in [0, \delta]} |V(t)|$ and $M = \max\{M_v, m_0 + 1\}$. Upon taking $\underline{u} : \bar{D}_\delta \to \mathbb{R}$ as $\underline{u}(x, t) = \underline{U}_{-n_j}^+(t)$ and $\bar{u} : \bar{D}_\delta \to \mathbb{R}$ as $\bar{u}(x, t) = V(t)$ as a regular subsolution and a regular supersolution to (B-R-D-C) with $f = \underline{f}'_{n_j}$ given by (8.22) (with M as above) and $u_0 = u_0^+$, then an application of Theorem 7.1 gives

$$\underline{U}_{n_j}^+(t) \leq V(t); \quad \forall t \in [0, \delta],$$

and so,

$$U(t) \leq V(t); \quad \forall t \in [0, \delta].$$

Therefore, it follows that $U : [0, \delta] \to \mathbb{R}$ is a minimal solution of (8.76), and so $U = \underline{U}^+$ on $\bar{D}_\delta$. Now, via (8.65) and (8.66), we have,

$$\underline{U}_{n_j}^+(t) = \lim_{x \to \infty} \underline{u}_{n_j}(x, t) = \liminf_{x \to \infty} \underline{u}_{n_j}(x, t) \leq \liminf_{x \to \infty} \underline{u}^c(x, t)$$

uniformly for $t \in [0, \delta]$ and $n_j \in \mathbb{N}$. It follows, via (8.70), that,

$$\liminf_{x \to \infty} \underline{u}^c(x, t) \geq U(t) = \underline{U}^+(t)$$

uniformly for $t \in [0, \delta]$, as required. The corresponding result as $x \to -\infty$ follows similarly. A symmetrical argument establishes the results for $\bar{u}^c(x, t)$ as $x \to \pm\infty$. $\square$

Remark 8.38 When the initial value problems,

$$U_t^\pm = f(U^\pm); \quad \forall t \in (0, T], \quad U^\pm(0) = u_0^\pm, \tag{8.77}$$

(corresponding to global versions of (8.63)) in Proposition 8.37 have unique solutions $U^\pm$ for $0 \leq T \leq T'$, then the maximal solutions $\bar{U}^\pm : [0, \delta] \to \mathbb{R}$ and the minimal solutions $\underline{U}^\pm : [0, \delta] \to \mathbb{R}$ to the initial value problems in Proposition 8.37 will be equal to $U^\pm : [0, \delta] \to \mathbb{R}$. It then follows from Proposition 8.37 and Definition 8.1 that

$$U^\pm(t) \leq \liminf_{x \to \pm\infty} \underline{u}^c(x, t) \leq \limsup_{x \to \pm\infty} \underline{u}^c(x, t)$$
$$\leq \liminf_{x \to \pm\infty} \bar{u}^c(x, t)$$
$$\leq \limsup_{x \to \pm\infty} \bar{u}^c(x, t) \leq U^\pm(t) \quad \text{uniformly for } t \in [0, \delta].$$

Thus, the following limits exist,

$$\lim_{x \to \pm\infty} \underline{u}^c(x, t) = \lim_{x \to \pm\infty} \overline{u}^c(x, t) = U^\pm(t) \quad \text{uniformly for } t \in [0, \delta]. \quad (8.78)$$

Suppose now that $\overline{u}^c, \underline{u}^c : \bar{D}_{T'} \to \mathbb{R}$ are the constructed maximal solution and the constructed minimal solution to (B-R-D-C) with f and u_0 on $\bar{D}_{T'}$, respectively. Since solutions to (8.77) are unique on $[0, T]$ for any $0 \leq T \leq T'$, then it follows that $[0, \delta]$ in (8.78) can be replaced by $[0, T']$.   ⌐

Finally, we provide the following conditional continuous dependence result.

Theorem 8.39 *Consider (B-R-D-C) with $f \in H_\alpha$ for some $\alpha \in (0, 1)$ and $u_0 \in BPC^2(\mathbb{R})$. For $u_0^* \in BPC^2(\mathbb{R})$ that satisfies*

$$\lim_{x \to \pm\infty} u_0^*(x) = u_0^\pm,$$

suppose that $u^* : \bar{D}_{T'} \to \mathbb{R}$ is the unique solution to (B-R-D-C) on $\bar{D}_T$ for any $0 < T \leq T'$. Moreover, suppose that the initial value problem

$$u_t = f(u); \quad \forall t \in (0, T'], \quad u(0) = c, \quad (8.79)$$

has solutions $U^\pm : [0, T'] \to \mathbb{R}$ for $c = u_0^+$ and $c = u_0^-$, which are unique on $[0, T]$ for any $0 \leq T \leq T'$. Then, for any $\epsilon > 0$, there exists $\delta^* > 0$ such that any solution $u : \bar{D}_{T'} \to \mathbb{R}$ to (B-R-D-C) with initial data $u_0 \in BPC^2(\mathbb{R})$ that satisfies $\|u_0 - u_0^*\|_B < \delta^*$ (of which there is at least one), also satisfies $\|u - u^*\|_A < \epsilon$.

Proof Since $u^* : \bar{D}_{T'} \to \mathbb{R}$ is the unique solution to (B-R-D-C) on $\bar{D}_T$ for any $0 < T \leq T'$, it follows that $u^* : \bar{D}_{T'} \to \mathbb{R}$ is a constructed solution, say in N_c applications of Theorem 8.3. It then follows from Proposition 8.37 and Remark 8.38 that

$$\lim_{x \to \pm\infty} u^*(x, t) = U^\pm(t) \quad (8.80)$$

uniformly for $t \in [0, T']$.

We now consider the first application of Theorem 8.3 in this construction procedure for $u^* : \bar{D}_{T'} \to \mathbb{R}$ and suppose that $\epsilon_1 > 0$. Let $\underline{u}_n, \overline{u}_n : \bar{D}_\delta \to \mathbb{R}$ be the unique solutions to (B-R-D-C)$_n^l$ and (B-R-D-C)$_n^u$ on $\bar{D}_\delta$, as employed in the proof of Theorem 8.3, respectively, with δ as in Theorem 8.3. Then, for any $X > 0$, via Corollary 8.20 and a symmetric argument, there exists $N_1 \in \mathbb{N}$ such that for all $n \geq N_1$,

$$\max\{|\underline{u}_n - u^*|, |\overline{u}_n - u^*|\} < \frac{\epsilon_1}{2} \quad \text{on } \bar{D}_\delta^{0, X}. \quad (8.81)$$

We now proceed with the argument for large positive x. It follows from Remark 8.18 and Proposition 8.16 (with a symmetrical argument) that

$$\underline{u}_n(x, t) \leq \underline{u}_{n+1}(x, t) \leq u^*(x, t) \leq \overline{u}_{n+1}(x, t) \leq \overline{u}(x, t); \quad \forall (x, t) \in \bar{D}_\delta. \tag{8.82}$$

Additionally, since $(B\text{-}R\text{-}D\text{-}C)_n^l$ and $(B\text{-}R\text{-}D\text{-}C)_n^u$ have $\underline{f}_n, \overline{f}_n \in L$, then it follows from [72, Theorem 5.2, p. 239] that

$$\lim_{x \to \infty} \underline{u}_n(x, t) = \underline{U}_n^+(t), \quad \lim_{x \to \infty} \overline{u}_n(x, t) = \overline{U}_n^+(t) \tag{8.83}$$

uniformly for $t \in [0, \delta]$, where $\underline{U}_n^+, \overline{U}_n^+ : [0, \delta] \to \mathbb{R}$ are respectively, the unique solutions to

$$\underline{U}_{nt}^+ = \underline{f}_n(\underline{U}_n^+) \quad \forall t \in (0, \delta], \quad \underline{U}_n^+(0) = u_0^+ - \frac{1}{2n}, \tag{8.84}$$

$$\overline{U}_{nt}^+ = \overline{f}_n(\overline{U}_n^+) \quad \forall t \in (0, \delta], \quad \overline{U}_n^+(0) = u_0^+ + \frac{1}{2n}. \tag{8.85}$$

Now, since $U^+ : [0, T'] \to \mathbb{R}$ is unique on $[0, \delta]$, it follows, as in the proof of Proposition 8.37, that there exists a subsequence $\{n_j\}_{j \in \mathbb{N}}$ of $1, 2, 3, \ldots,$ such that

$$\underline{U}_{n_j}^+ \to U^+ \text{ and } \overline{U}_{n_j}^+ \to U^+ \text{ uniformly as } n_j \to \infty \text{ on } [0, \delta]. \tag{8.86}$$

Thus, it follows from (8.86) and (8.82) that there exists $N_1' \in \mathbb{N}$ such that for all $n \geq N_1'$,

$$\max\{|(\underline{U}_n^+ - U^+)(t)|, |(\overline{U}_n^+ - U^+)(t)|\} < \frac{\epsilon_1}{2}; \quad \forall t \in [0, \delta]. \tag{8.87}$$

Now, it follows from (8.80), (8.82), (8.83) and (8.87) that there exists $X_1 > 0$ such that for all $n \geq N_1'$,

$$-\frac{\epsilon_1}{2} < (\underline{u}_n - u^*)(x, t) \leq (\overline{u}_n - u^*)(x, t) < \frac{\epsilon_1}{2}; \quad \forall (x, t) \in [X_1, \infty) \times [0, \delta]. \tag{8.88}$$

Via a symmetrical argument, it follows that there exists $X_1' > 0$ and $N_1'' \in \mathbb{N}$ such that for all $n \geq N_1''$,

$$-\epsilon_1 < (\underline{u}_n - u^*)(x, t) \leq (\overline{u}_n - u^*)(x, t) < \epsilon_1; \quad \forall (x, t) \in \bar{D}_\delta \backslash \bar{D}_\delta^{0, X_1'}. \tag{8.89}$$

Therefore, it follows from (8.81) with $X = X_1'$ and (8.89) that there exists $N_1* \in \mathbb{N}$ such that for all $n \geq N_1*$, we have

$$-\epsilon_1 < (\underline{u}_n - u^*)(x, t) \leq (\overline{u}_n - u^*)(x, t) < \epsilon_1; \quad \forall (x, t) \in \bar{D}_\delta. \tag{8.90}$$

Now, set $\delta_1 = 1/(2N_1*)$ and suppose that $u : \bar{D}_\delta \to \mathbb{R}$ is a solution to (B-R-D-C) with f and $u_0 \in \mathrm{BPC}^2(\mathbb{R})$ on $\bar{D}_\delta$, such that $\|u_0 - u_0^*\|_B < \delta_1$. Upon setting $M = \max\{\|u\|_A, m_0 + 1\}$, and taking $\underline{u} : \bar{D}_\delta \to \mathbb{R}$ as $\underline{u}(x, t) = \underline{u}_{N_1*}(x, t)$ and $\bar{u} : \bar{D}_\delta \to \mathbb{R}$ as $\bar{u}(x, t) = u(x, t)$ as a regular subsolution and a regular supersolution to (B-R-D-C) with $\underline{f}'_{N_1*}$ given by (8.22) (with M above) and u_0, it follows from Theorem 7.1 that

$$\underline{u}_{N_1*}(x, t) \le u(x, t); \quad \forall (x, t) \in \bar{D}_\delta. \tag{8.91}$$

A symmetrical argument then establishes, with (8.91) that

$$\underline{u}_{N_1*}(x, t) \le u(x, t) \le \bar{u}_{N_1*}(x, t); \quad \forall (x, t) \in \bar{D}_\delta. \tag{8.92}$$

It follows from (8.92) that (B-R-D-C) with f and u_0 is a priori bounded on $\bar{D}_\delta$, and hence, that there exists a constructed maximal (minimal) solution $\bar{u}^c(\underline{u}^c) : \bar{D}_\delta \to \mathbb{R}$ to (B-R-D-C) with f and u_0, and additionally, via (8.90), that

$$\|u - u^*\|_A < \epsilon_1. \tag{8.93}$$

Therefore, we have exhibited that for any $\epsilon_1 > 0$, there exists $\delta_1 > 0$ such that (B-R-D-C) with f and u_0 such that $\|u_0 - u_0^*\|_B < \delta_1$, has a constructed maximal (minimal) solution on $\bar{D}_\delta$ and for any solution u to (B-R-D-C) with f and u_0 on $\bar{D}_\delta$, then $\|u - u^*\|_A < \epsilon_1$.

We now construct the solution $u^* : \bar{D}_{T'} \to \mathbb{R}$ to (B-R-D-C) with f and u_0^* in the supposed N_c steps, generating for any $\epsilon_i \in \mathbb{R}$, the pair $(\delta_i, \epsilon_i) \in \mathbb{R}^2$ for each $i = 1, \ldots, N_c$, as produced in the above construction. By setting

$$\epsilon_{N_c} = \epsilon \text{ and } \epsilon_i = \min\{\delta_{i+1}, \epsilon\} \text{ for } i = 1, \ldots, N_c - 1,$$

we obtain $\delta_1 := \delta^*$, such that, for all $u_0 \in \mathrm{BPC}^2(\mathbb{R})$ such that $\|u_0 - u_0^*\|_B < \delta^*$, there exist a constructed maximal (minimal) solution $\bar{u}^c(\underline{u}^c) : \bar{D}_{T'} \to \mathbb{R}$ to (B-R-D-C) with f and u_0, and moreover, that $\|u - u^*\|_A < \epsilon$, as required. $\qquad \square$

Remark 8.40 In Theorem 8.39, there is no guarantee that the solution to (B-R-D-C) with f and u_0 such that $\|u_0 - u^*\|_B < \delta^*$ is unique. The result is an extension of a corresponding result for the initial value problem for a first order ordinary differential equation, as detailed in Chapter 2, Section 4 of [17].

9

Application to Specific Problems

In this chapter we apply results contained in Chapter 7 and Chapter 8 to the (B-R-D-C) problem given by (1.7)–(1.9), the problem studied in [5] and [48], and the problem given by (1.16)–(1.18). Specifically, for the first problem, we establish uniform global well-posedness on $\mathrm{BPC}^2_+(\mathbb{R})$. For the second problem, we establish global well-posedness on $\mathrm{BPC}^2_{+\prime}(\mathbb{R})$, and under additional technical conditions obtain a uniform global well-posedness result on a strict subset of $\mathrm{BPC}^2_{+\prime}(\mathbb{R})$. For the third problem, we establish a uniform global well-posedness result on $\mathrm{BPC}^2_{+\prime}(\mathbb{R})$. We also exhibit several distinctive qualitative properties of solutions to these problems.

9.1 $f(u) = -[u^p]^+$

We consider first the (B-R-D-C) problem with reaction function $f : \mathbb{R} \to \mathbb{R}$ given by

$$f(u) = -[u^p]^+ = \begin{cases} -u^p & ; u \geq 0 \\ 0 & ; u < 0, \end{cases} \tag{9.1}$$

where $p \in (0, 1)$. We restrict attention to initial data which is non-negative, that is $u_0 \in \mathrm{BPC}^2_+(\mathbb{R})$, which is the situation of interest in modelling problems arising from chemical kinetics, biology and combustion. Throughout the subsection this (B-R-D-C) will be referred to as (S-R-D-C-1). It should be mentioned that the problem (S-R-D-C-1) has been considered in [27] and [25]. The following section is included as an illustration of how the problem specific results obtained in these papers follow directly as an application of the generic theory developed in Chapter 7 and Chapter 8. To begin, we require,

Proposition 9.1 *The reaction function* $f : \mathbb{R} \to \mathbb{R}$ *given by* (9.1) *satisfies* $f \in H_p \cap L_u$, *with Hölder constant* $k_H = 1$ *and upper Lipschitz constant which can be taken as any positive real number, on any closed bounded interval* $E \subset \mathbb{R}$.

Proof First, we establish $f \in H_p$. Let $x, y \in \mathbb{R}$. When $0 \le x \le y$ then

$$|f(y) - f(x)| = |-y^p + x^p| = y^p - x^p \le (y - x)^p = |y - x|^p. \quad (9.2)$$

Next, when $x \le y \le 0$, then

$$|f(y) - f(x)| = 0 \le |y - x|^p. \quad (9.3)$$

Finally, with $x < 0 < y$, then

$$|f(y) - f(x)| = y^p < (y - x)^p = |y - x|^p. \quad (9.4)$$

It follows from (9.2), (9.3) and (9.4) that $f \in H_p$ and $k_H = 1$ is a Hölder constant for f on any closed bounded interval $E \subset \mathbb{R}$. Finally we observe from (9.1) that f is non-increasing on $\mathbb{R}$. It follows from Proposition 2.7 that $f \in L_u$, and that, on any closed bounded interval $E \subset \mathbb{R}$, any $k_u > 0$ serves as an upper Lipschitz constant for f. $\qquad \square$

We next establish that (S-R-D-C-1) is a priori bounded on $\bar{D}_T$ for any $T > 0$.

Proposition 9.2 *Let* $u : \bar{D}_T \to \mathbb{R}$ *be any solution to* (S-R-D-C-1) *with initial data* $u_0 \in BPC_+^2(\mathbb{R})$, *then*

$$0 \le u(x, t) \le M_0, \quad \forall (x, t) \in \bar{D}_T,$$

with $M_0 = \sup_{\lambda \in \mathbb{R}} \{ u_0(\lambda) \} \ge 0$.

Proof We begin by defining the functions $\bar{u}, \underline{u} : \bar{D}_T \to \mathbb{R}$ to be

$$\bar{u}(x, t) = M_0, \quad (9.5)$$
$$\underline{u}(x, t) = 0 \quad (9.6)$$

for all $(x, t) \in \bar{D}_T$. Observe, via (9.5) and (9.6), that the following inequalities hold, namely,

$$\bar{u}_t - \bar{u}_{xx} - f(\bar{u}) \ge 0 \quad \text{on } D_T, \quad (9.7)$$
$$\underline{u}_t - \underline{u}_{xx} - f(\underline{u}) \le 0 \quad \text{on } D_T, \quad (9.8)$$
$$\underline{u}(x, 0) \le u(x, 0) \le \bar{u}(x, 0) \quad \forall x \in \mathbb{R} \quad (9.9)$$

for f given by (9.1). It follows that $\underline{u}$ is a regular subsolution and $\bar{u}$ is a regular supersolution to (S-R-D-C-1). Thus, via Comparison Theorem 7.1 we have

$$0 \le u(x,t) \le M_0 \tag{9.10}$$

for all $(x,t) \in \bar{D}_T$, as required. $\square$

Now, we have the first main result of this subsection, namely,

Theorem† 9.3 *The problem (S-R-D-C-1) has a unique global solution on $\bar{D}_\infty$.*

Proof Proposition 9.1 establishes that $f \in H_p \cap L_u$, whilst Proposition 9.2 establishes that (S-R-D-C-1) is a priori bounded on $\bar{D}_T$ for any $T > 0$. Now, uniqueness follows from Theorem 7.2 and global existence follows from Theorem 8.25. $\square$

Moreover, we have,

Theorem† 9.4 *The problem (S-R-D-C-1) is globally well-posed on $BPC_+^2(\mathbb{R})$.*

Proof Follows from Theorem 9.3, Proposition 9.1 and Corollary 7.8. $\square$

We now establish qualitative properties of the solutions to (S-R-D-C-1).

Proposition† 9.5 *Let $u : \bar{D}_\infty \to \mathbb{R}$ be the unique solution to (S-R-D-C-1) with $u_0 \in BPC_+^2(\mathbb{R})$. Then, for given $0 < p < 1$, there exists $0 \le t_c \le \frac{M_0^{(1-p)}}{(1-p)}$ depending only upon $M_0 = \sup_{\lambda \in \mathbb{R}}\{u_0(\lambda)\}$ such that*

$$u(x,t) = 0; \quad \forall (x,t) \in \bar{D}_\infty^{t_c}.$$

Proof Let $u : \bar{D}_\infty \to \mathbb{R}$ be the unique solution of (S-R-D-C-1). Now consider the function $z : \bar{D}_\infty \to \mathbb{R}$ defined to be

$$z(x,t) = \begin{cases} \left(M_0^{(1-p)} - (1-p)t\right)^{\frac{1}{(1-p)}} & ; (x,t) \in \bar{D}_{t_c^0} \\ 0 & ; (x,t) \in \bar{D}_\infty^{t_c^0}, \end{cases} \tag{9.11}$$

where $t_c^0 = (1-p)^{-1} M_0^{(1-p)}$. It is readily verified that z is continuous on $\bar{D}_\infty$, z is bounded as $|x| \to \infty$ uniformly for all $t \in (0,\infty)$, and that z_t and z_{xx} exist and are continuous on D_∞. Moreover, it is readily verified that z is a regular supersolution to (S-R-D-C-1) on $\bar{D}_\infty$. Similarly, it is readily verified that the zero function on $\bar{D}_\infty$ is a regular subsolution to (S-R-D-C-1) on $\bar{D}_\infty$. An application of Comparison Theorem 7.1 gives

$$0 \le u(x, t) \le z(x, t); \quad \forall (x, t) \in \bar{D}_\infty,$$

and the result follows. □

We are now in a position to state a uniform global continuous dependence result for (S-R-D-C-1),

Theorem† 9.6 *Let* $u^* : \bar{D}_\infty \to \mathbb{R}$ *be the unique solution to (S-R-D-C-1) corresponding to* $u_0^* \in BPC_+^2(\mathbb{R})$. *Let* $u : \bar{D}_\infty \to \mathbb{R}$ *be the unique solution to (S-R-D-C-1) corresponding to* $u_0 \in BPC_+^2(\mathbb{R})$. *Then, given any* $\epsilon > 0$, *there exists* $\delta > 0$, *depending only upon* $\|u_0^*\|_B$ *and* ϵ, *such that for all* $u_0 \in BPC_+^2(\mathbb{R})$ *with*

$$\|(u_0 - u_0^*)\|_B < \delta,$$

then

$$\|(u - u^*)(\cdot, t)\|_B < \epsilon$$

for all $t \in [0, \infty)$, *and so (S-R-D-C-1) is uniformly globally well-posed on* $BPC_+^2(\mathbb{R})$.

Proof Let $M^* = \|u_0^*\|_B + 1$ and $T^* = (1 - p)^{-1} M^{*(1-p)}$. Now, given $\epsilon > 0$, it follows from Theorem 9.4, that there exists $\delta' > 0$ such that, for all $u_0 \in BPC_+^2(\mathbb{R})$ with $\|(u_0 - u_0^*)\|_B < \delta'$, then $\|(u - u^*)\|_A < \epsilon$ on $\bar{D}_{T^*}$, and δ' depending only upon ϵ and $\|u_0^*\|_B$. Now set $\delta = \min(1, \delta')$, so that, for any $u_0 \in BPC_+^2(\mathbb{R})$ with $\|(u_0 - u_0^*)\|_B < \delta$, then

$$\|u_0\|_B < \|u_0^*\|_B + \delta \le \|u_0^*\|_B + 1. \tag{9.12}$$

It then follows from Proposition 9.5 and (9.12), that

$$u_0^*(x, t) = u(x, t) = 0; \quad \forall (x, t) \in \bar{D}_\infty^{T^*}.$$

Therefore, we have that for all $u_0 \in BPC_+^2(\mathbb{R})$ with $\|(u_0 - u_0^*)\|_B < \delta$, then $\|(u - u^*)(\cdot, t)\|_B < \epsilon$ for all $t \in [0, T]$ and any $T > 0$, with $\delta > 0$ depending only upon ϵ and $\|u_0^*\|_B$. It follows that (S-R-D-C-1) is uniformly globally well-posed on $BPC_+^2(\mathbb{R})$, via Theorem 9.3. □

Remark 9.7 In fact, Theorem 9.6 establishes that for every $u_0^* \in BPC_+^2(\mathbb{R})$, the corresponding unique global solution $u^* : \bar{D}_\infty \to \mathbb{R}$ is Liapunov stable with respect to perturbations in initial data $u_0^* + \delta u_0 \in BPC_+^2(\mathbb{R})$. Moreover, it then follows from Proposition 9.5, that for every $u_0^* \in BPC_+^2(\mathbb{R})$ the corresponding unique global solution $u^* : \bar{D}_\infty \to \mathbb{R}$ is asymptotically stable with respect to perturbations in initial data $u_0^* + \delta u_0 \in \mathrm{BPC}_+^2(\mathbb{R})$. ⌙

We next consider further the qualitative properties of the solution to (S-R-D-C-1). To this end we observe that the non-negative functions $u_L : \mathbb{R} \to \mathbb{R}$ and $u_R : \mathbb{R} \to \mathbb{R}$, given by

$$u_L(x) = \begin{cases} \left(\frac{(1-p)}{2} \sqrt{\frac{2}{(1+p)}} \right)^{2/(1-p)} (x_0 + x)^{2/(1-p)} & ; x \geq -x_0 \\ 0 & ; x < -x_0 \end{cases}$$

$$u_R(x) = \begin{cases} \left(\frac{(1-p)}{2} \sqrt{\frac{2}{(1+p)}} \right)^{2/(1-p)} (x_0 - x)^{2/(1-p)} & ; x \leq x_0 \\ 0 & ; x > x_0 \end{cases}$$

for any fixed $x_0 \geq 0$, are both solutions of the ordinary differential equation

$$u_{xx} - [u^p]^+ = 0; \quad -\infty < x < \infty.$$

Now consider (S-R-D-C-1) when $\mathrm{supp}_{x \in \mathbb{R}} u_0(x)$ is bounded. We can then choose x_0 sufficiently large (depending upon $\mathrm{supp}_{x \in \mathbb{R}} u_0(x)$ and $\|u_0\|_B$) so that

$$0 \leq u_0(x) \leq \min\{u_L(x), u_R(x)\}; \quad \forall x \in \mathbb{R}. \tag{9.13}$$

Now introduce $w_L, w_R : \bar{D}_T \to \mathbb{R}$ as

$$w_L(x,t) = u(x,t) - u_L(x), \quad w_R(x,t) = u(x,t) - u_R(x)$$

for all $(x,t) \in \bar{D}_T$, with $u : \bar{D}_\infty \to \mathbb{R}$ being the solution to (S-R-D-C-1) and x_0 chosen to satisfy (9.13). It then follows from Theorem 4.4 in [49] that

$$w_L(x,t) \leq 0, \quad w_R(x,t) \leq 0$$

for all $(x,t) \in \bar{D}_T$. Therefore, it follows from Proposition 9.2 and the above, that

$$0 \leq u(x,t) \leq \min\{u_L(x), u_R(x)\}; \quad \forall (x,t) \in \bar{D}_\infty. \tag{9.14}$$

We therefore have,

Theorem† 9.8 *Let* $u : \bar{D}_\infty \to \mathbb{R}$ *be the unique solution to (S-R-D-C-1) when* $\mathrm{supp}_{x \in \mathbb{R}} u_0(x)$ *is bounded. Then, there exists* $x_0 \geq 0$ *(depending upon* $\mathrm{supp}_{x \in \mathbb{R}} u_0(x)$ *and* $\|u_0\|_B$*) such that*

$$\mathrm{supp}_{(x,t) \in \bar{D}_\infty} u(x,t) \subseteq [-x_0, x_0] \times [0, (1-p)^{-1} \|u_0\|_B^{(1-p)}].$$

Proof Follows from (9.13) and (9.14) together with Proposition 9.5. □

To conclude this section we remark that global well-posedness results can be developed, in general, when $f \in H_\alpha \cap L_u$ and f is non-increasing, with details given in [50].

9.2 $f(u) = [u^p]^+$

In this section we consider the (B-R-D-C) problem when the reaction function $f : \mathbb{R} \to \mathbb{R}$ is given by

$$f(u) = [u^p]^+ = \begin{cases} u^p & ; u \geq 0 \\ 0 & ; u < 0, \end{cases} \tag{9.15}$$

where $p \in (0, 1)$. In particular, we restrict attention to initial data $u_0 \in \mathrm{BPC}^2_{+'}(\mathbb{R})$, with

$$\sup_{x \in \mathbb{R}} \{u_0(x)\} = M_0 > 0, \quad \inf_{x \in \mathbb{R}} \{u_0(x)\} = m_0 \geq 0. \tag{9.16}$$

The aim of this section is to provide a global well-posedness result for this specific (B-R-D-C). Bearing this in mind, the technical condition (9.16) is shown to be necessary by Example 8.28. Throughout the section, this (B-R-D-C) will be referred to as (S-R-D-C-2).

It should be mentioned that the problem (S-R-D-C-2) was studied in [5], [48] and [54]. A more generic approach is adopted here, utilizing the general results of Chapter 6 and Chapter 8.

Proposition 9.9 *The reaction function $f : \mathbb{R} \to \mathbb{R}$ given by (9.15) is such that $f \in H_p$, with Hölder constant $k_H = 1$ on any closed bounded interval $E \subset \mathbb{R}$.*

Proof Follows from Proposition 9.1.   □

Remark 9.10 The reaction function $f : \mathbb{R} \to \mathbb{R}$ given by (9.15) is such that $f \notin L_u$. This follows since for $u > 0$, then

$$\frac{(f(u) - f(0))}{(u - 0)} = u^{p-1},$$

which is not bounded above as $u \to 0^+$.   ⌋

We now establish a priori bounds for (S-R-D-C-2). We begin with,

Proposition 9.11 *Let $u : \bar{D}_T \to \mathbb{R}$ be any solution to (S-R-D-C-2). Then,*

$$m_0 \le u(x,t) \le \left((1-p)t + M_0^{(1-p)}\right)^{1/(1-p)}; \quad \forall (x,t) \in \bar{D}_T.$$

Proof We begin with the right hand inequality. First we introduce $\bar{f} : \mathbb{R} \to \mathbb{R}$ such that

$$\bar{f}(u) = \begin{cases} u^p & ; u \ge M_0 \\ M_0^p & ; u < M_0 \end{cases}$$

and we observe that

$$w_t - w_{xx} - \bar{f}(w) = 0 \quad \text{on } D_T, \tag{9.17}$$

where $w : \bar{D}_T \to \mathbb{R}$ is such that

$$w(x,t) = \left((1-p)t + M_0^{(1-p)}\right)^{1/(1-p)}; \quad \forall (x,t) \in \bar{D}_T.$$

Moreover,

$$u_t - u_{xx} - \bar{f}(u) = f(u) - \bar{f}(u) \le 0 \quad \text{on } D_T. \tag{9.18}$$

Since $\bar{f} \in L_u$, it follows from Comparison Theorem 7.1, via (9.17) and (9.18), that

$$u(x,t) \le \left((1-p)t + M_0^{(1-p)}\right)^{1/(1-p)}; \quad \forall (x,t) \in \bar{D}_T.$$

Next we observe that $(u - m_0)$ satisfies

$$(u - m_0)_t - (u - m_0)_{xx} \ge 0 \quad \text{on } D_T.$$

It then follows from the Maximum Principle given by Theorem 3.6 that

$$u(x,t) \ge m_0; \quad \forall (x,t) \in \bar{D}_T.$$

This completes the proof. $\qquad \square$

Thus we have established, via Proposition 9.11, that (S-R-D-C-2) is a priori bounded on $\bar{D}_T$ for any $T > 0$. We therefore have,

Theorem$_\ddagger$ 9.12 *(S-R-D-C-2) has a global constructed maximal solution $\bar{u}^c : \bar{D}_\infty \to \mathbb{R}$ and a global constructed minimal solution $\underline{u}^c : \bar{D}_\infty \to \mathbb{R}$. Moreover,*

$$m_0 \le \underline{u}^c(x,t) \le \bar{u}^c(x,t) \le \left((1-p)t + M_0^{(1-p)}\right)^{1/(1-p)}; \quad \forall (x,t) \in \bar{D}_\infty.$$

Proof This follows from Proposition 9.9 and Proposition 9.11 together with Theorem 8.25. ☐

In fact, we have,

Corollary‡ 9.13 *Let* $\overline{u}^c : \bar{D}_\infty \to \mathbb{R}$ *be the global constructed maximal solution to (S-R-D-C-2). Then,*

$$\left((1-p)t + m_0^{(1-p)} \right)^{1/(1-p)}$$

$$\leq \overline{u}^c(x,t) \leq \left((1-p)t + M_0^{(1-p)} \right)^{1/(1-p)}; \quad \forall (x,t) \in \bar{D}_\infty.$$

Proof This follows from Theorem 9.12 and Theorem 8.26. ☐

We now refer to a result given in Aguirre and Escobedo [5], which will be used to obtain the following uniqueness result for (S-R-D-C-2). The proof has been omitted due to the uniqueness result in the following section being obtained by a similar approach.

Theorem 9.14 (Aguirre and Escobedo) *Let* $\underline{w} : \bar{D}_T \to \mathbb{R}$ *and* $\overline{w} : \bar{D}_T \to \mathbb{R}$ *be a non-negative regular subsolution and a non-negative regular supersolution to (S-R-D-C-2), respectively, with* $\overline{w}(\cdot, 0) \in BPC_{+}^2(\mathbb{R})$. *Then,*

$$\underline{w}(x,t) \leq \overline{w}(x,t); \quad \forall (x,t) \in \bar{D}_T.$$

Remark 9.15 Observe that the condition $\overline{w}(\cdot, 0) \in BPC_{+}^2(\mathbb{R})$ in Theorem 9.14 is crucial, otherwise Theorem 9.14 and Proposition 9.11 would establish uniqueness for (S-R-D-C-2) when $u_0(x) = 0$ for all $x \in \mathbb{R}$, which is false (see Example 8.28). ⌐

However, we now have,

Theorem† 9.16 *(S-R-D-C-2) has a unique global solution* $u : \bar{D}_\infty \to \mathbb{R}$. *Moreover,*

$$\left((1-p)t + m_0^{(1-p)} \right)^{1/(1-p)}$$

$$\leq u(x,t) \leq \left((1-p)t + M_0^{(1-p)} \right)^{1/(1-p)}; \quad \forall (x,t) \in \bar{D}_\infty.$$

Proof Let $u_1 : \bar{D}_\infty \to \mathbb{R}$ and $u_2 : \bar{D}_\infty \to \mathbb{R}$ be global solutions to (S-R-D-C-2). Since $u_0 \in BPC_{+}^2(\mathbb{R})$ satisfies (9.16), then Proposition 9.11 and Comparison Theorem 9.14 give

$$u_1(x, t) \le u_2(x, t) \le u_1(x, t); \quad \forall (x, t) \in \bar{D}_\infty$$

and so $u_1 = u_2$ on $\bar{D}_\infty$, and uniqueness is established. Existence and the inequalities follow via Theorem 9.12 and Corollary 9.13.   □

We next consider continuous dependence for (S-R-D-C-2). This result has been obtained via an alternative approach in [5].

Theorem† 9.17 *Let $u^* : \bar{D}_\infty \to \mathbb{R}$ and $u : \bar{D}_\infty \to \mathbb{R}$ be the unique global solutions to (S-R-D-C-2) with initial data $u_0^* \in BPC_{+'}^2(\mathbb{R})$ and $u_0 \in BPC_{+'}^2(\mathbb{R})$ respectively. Then, given any $T > 0$ and any $\epsilon > 0$, there exists a constant $\delta > 0$ (depending only upon ϵ, u_0^*, T and p) such that whenever*

$$||u_0 - u_0^*||_B < \delta,$$

then

$$||(u - u^*)(\cdot, t)||_B < \epsilon; \quad \forall t \in [0, T].$$

Proof Let $\delta > 0$ and consider $||u_0 - u_0^*||_B < \delta$. It follows from Lemma 5.10 and Proposition 9.9 that

$$|(u^* - u)(x, t)|$$

$$\le \frac{1}{\sqrt{\pi}} \int_{-\infty}^{\infty} |(u_0^* - u_0)(x + 2\sqrt{t}\lambda)| e^{-\lambda^2} d\lambda$$

$$+ \frac{1}{\sqrt{\pi}} \int_0^t \int_{-\infty}^{\infty} |(f(u^*) - f(u))(x + 2\sqrt{t - \tau}\lambda, \tau)| e^{-\lambda^2} d\lambda d\tau \quad (9.19)$$

$$\le \frac{1}{\sqrt{\pi}} \int_{-\infty}^{\infty} \delta e^{-\lambda^2} d\lambda$$

$$+ \frac{1}{\sqrt{\pi}} \int_0^t \int_{-\infty}^{\infty} |(u^* - u)(x + 2\sqrt{t - \tau}\lambda, \tau)|^p e^{-\lambda^2} d\lambda d\tau$$

$$\le \delta + \frac{1}{\sqrt{\pi}} \int_0^t \int_{-\infty}^{\infty} ||(u^* - u)(\cdot, \tau)||_B^p e^{-\lambda^2} d\lambda d\tau$$

$$= \delta + \int_0^t ||(u^* - u)(\cdot, \tau)||_B^p d\tau \quad (9.20)$$

for all $(x, t) \in \bar{D}_T$, from which it follows that

$$||(u^* - u)(\cdot, t)||_B \le \delta + \int_0^t ||(u^* - u)(\cdot, \tau)||_B^p d\tau \quad (9.21)$$

for all $t \in [0, T]$. Now, define $F : [0, T] \to \mathbb{R}$ to be

$$F(t) = \delta + \int_0^t \|(u^* - u)(\cdot, \tau)\|_B^p \, d\tau; \quad \forall t \in [0, T]. \tag{9.22}$$

It follows from (9.22), Corollary 5.16 and the fundamental theorem of calculus, that F is differentiable on $[0, T]$ and it follows from (9.21) that F satisfies the following differential inequality,

$$\frac{1}{(F(\tau))^p} \frac{dF(\tau)}{d\tau} \le 1; \quad \forall \tau \in [0, T]. \tag{9.23}$$

Upon integrating both sides of (9.23), it follows that

$$F(t)^{(1-p)} \le F(0)^{(1-p)} + t(1-p) = \delta^{(1-p)} + t(1-p); \quad \forall t \in [0, T]. \tag{9.24}$$

It now follows from (9.24), that

$$F(t) \le \left(\delta^{(1-p)} + (1-p)t\right)^{1/(1-p)}, \quad \forall t \in [0, T] \tag{9.25}$$

and so, from (9.25) and (9.21), we have

$$\|(u^* - u)(\cdot, t)\|_B \le \left(\delta^{(1-p)} + (1-p)t\right)^{1/(1-p)}; \quad \forall t \in [0, T]. \tag{9.26}$$

Now choose δ sufficiently small so that $T_1 = (1-p)^{-1}\delta^{(1-p)} < T$. Then, via (9.26),

$$\|(u^*-u)(\cdot, t)\|_B \le \left(\delta^{(1-p)} + (1-p)T_1\right)^{1/(1-p)} = 2^{1/(1-p)}\delta; \quad \forall t \in [0, T_1]. \tag{9.27}$$

Now, $u : \bar{D}_T \to \mathbb{R}$ and $u^* : \bar{D}_T \to \mathbb{R}$ both satisfy the lower bound in Theorem 9.16, with $m_0 = 0$. It then follows (since f given by (9.15) is differentiable on $(0, \infty)$) via the mean value theorem, that for all $(s, \tau) \in D_T$ there exists $\theta \ge ((1-p)\tau)^{1/(1-p)}$ such that

$$\begin{aligned}|(f(u^*) - f(u))(s, \tau)| &\le |f'(\theta)| \, |(u^* - u)(s, \tau)| \\ &\le p\theta^{(p-1)} |(u^* - u)(s, \tau)| \\ &\le p((1-p)\tau)^{(p-1)/(1-p)} |(u^* - u)(s, \tau)| \\ &\le \frac{p}{(1-p)\tau}\|(u^* - u)(\cdot, \tau)\|_B. \end{aligned} \tag{9.28}$$

Via Proposition 9.9, $f \in H_p$ with Hölder constant $k_H = 1$ on any closed bounded interval and so, upon substituting (9.27) and (9.28) into (9.19), we obtain

$|(u^* - u)(x,t)|$

$$\leq \delta + \frac{1}{\sqrt{\pi}} \int_0^{T_1} \int_{-\infty}^{\infty} \left|(f(u^*) - f(u))\left(x + 2\sqrt{t-\tau}\lambda, \tau\right)\right| e^{-\lambda^2} d\lambda d\tau$$

$$+ \frac{1}{\sqrt{\pi}} \int_{T_1}^{t} \int_{-\infty}^{\infty} \left|(f(u^*) - f(u))\left(x + 2\sqrt{t-\tau}\lambda, \tau\right)\right| e^{-\lambda^2} d\lambda d\tau$$

$$\leq \delta + \frac{1}{\sqrt{\pi}} \int_0^{T_1} \int_{-\infty}^{\infty} ||(u^* - u)(\cdot, \tau)||_B^p \, e^{-\lambda^2} d\lambda d\tau$$

$$+ \frac{1}{\sqrt{\pi}} \int_{T_1}^{t} \int_{-\infty}^{\infty} \frac{p}{(1-p)\tau} ||(u^* - u)(\cdot, \tau)||_B e^{-\lambda^2} d\lambda d\tau$$

$$\leq \delta + \int_0^{T_1} 2^{p/(1-p)} \delta^p d\tau + \int_{T_1}^{t} \frac{p}{(1-p)\tau} ||(u^* - u)(\cdot, \tau)||_B d\tau$$

$$= \left(1 + \frac{2^{p/(1-p)}}{(1-p)}\right) \delta + \int_{T_1}^{t} \frac{p}{(1-p)\tau} ||(u^* - u)(\cdot, \tau)||_B d\tau \tag{9.29}$$

for all $(x,t) \in \bar{D}_T^{T_1}$, from which it follows,

$$||(u^* - u)(\cdot, t)||_B \leq \left(1 + \frac{2^{p/(1-p)}}{(1-p)}\right) \delta$$

$$+ \int_{T_1}^{t} \frac{p}{(1-p)\tau} ||(u^* - u)(\cdot, \tau)||_B d\tau; \quad \forall t \in [T_1, T]. \tag{9.30}$$

Now define $G : [T_1, T] \to \mathbb{R}^+$ to be

$$G(t) = \left(1 + \frac{2^{p/(1-p)}}{(1-p)}\right) \delta + \int_{T_1}^{t} \frac{p}{(1-p)\tau} ||(u^* - u)(\cdot, \tau)||_B d\tau; \quad \forall t \in [T_1, T]. \tag{9.31}$$

It follows from (9.30), (9.31), Corollary 5.16 and the fundamental theorem of calculus, that G is differentiable on $[T_1, T]$ and satisfies

$$\left(\frac{1}{G(\tau)}\right) \frac{dG(\tau)}{d\tau} \leq \frac{p}{(1-p)\tau}; \quad \forall \tau \in [T_1, T]. \tag{9.32}$$

Upon integrating both sides of (9.32) with respect to τ from T_1 to $t \in [T_1, T]$, we obtain

$$\ln\left(\frac{G(t)}{\left(1 + \frac{2^{p/(1-p)}}{(1-p)}\right)\delta}\right) \leq \frac{p}{(1-p)} \ln\left(\frac{t(1-p)}{\delta^{(1-p)}}\right) \leq \ln\left(\frac{(T(1-p))^{p/(1-p)}}{\delta^p}\right) \tag{9.33}$$

for all $t \in [T_1, T]$. Taking exponentials of both sides of (9.33) and re-arranging gives

$$G(t) \leq (T(1-p))^{p/(1-p)} \left(1 + \frac{2^{p/(1-p)}}{(1-p)}\right) \delta^{(1-p)}; \quad \forall t \in [T_1, T]. \quad (9.34)$$

Combining (9.27) and (9.34) gives the following bound,

$$||(u^*-u)(\cdot, t)||_B \leq \begin{cases} 2^{1/(1-p)}\delta & ; t \in [0, T_1] \\ (T(1-p))^{p/(1-p)} \left(1 + \frac{2^{p/(1-p)}}{(1-p)}\right) \delta^{(1-p)} & ; t \in [T_1, T]. \end{cases}$$
$$(9.35)$$

Therefore, given any $\epsilon > 0$, take

$$\delta = \min \left\{ \frac{\epsilon}{2^{1+1/(1-p)}}, \left(\frac{\epsilon}{2(T(1-p))^{p/(1-p)} \left(1 + \frac{2^{p/(1-p)}}{(1-p)}\right)}\right)^{1/(1-p)}, \right.$$
$$\left. \left(\frac{T}{2(1-p)}\right)^{1/(1-p)} \right\}.$$

Then, via (9.35),

$$||(u^* - u)(\cdot, t)||_B \leq \epsilon/2 < \epsilon$$

for all $t \in [0, T]$, as required. $\qquad \square$

Corollary† 9.18 *(S-R-D-C-2) is globally well-posed on* $BPC^2_{+'}(\mathbb{R})$.

Proof Follows directly from Theorem 9.16 and Theorem 9.17. $\qquad \square$

Remark 9.19 Although (S-R-D-C-2) is globally well-posed on $BPC^2_{+'}(\mathbb{R})$, it is not uniformly globally well-posed on $BPC^2_{+'}(\mathbb{R})$. To illustrate this consider $u_1, u_2 : \bar{D}_\infty \to \mathbb{R}$ defined to be

$$u_1(x, t) = \left(M^{(1-p)} + (1-p)t\right)^{1/(1-p)}$$

and

$$u_2(x, t) = \left((M+\delta)^{(1-p)} + (1-p)t\right)^{1/(1-p)}$$

for all $(x, t) \in \bar{D}_\infty$, where $M, \delta > 0$. It is readily verified that u_1 and u_2 are solutions to (S-R-D-C-2) with initial data $u_0^1, u_0^2 \in BPC^2_{+'}(\mathbb{R})$ given by

$$u_0^1(x) = M \quad \text{and} \quad u_0^2(x) = M + \delta$$

for all $x \in \mathbb{R}$. However, for any $\delta > 0$,

$$||(u_2 - u_1)(\cdot, t)||_B \to \infty \text{ as } t \to \infty. \qquad \lrcorner$$

Although a uniform global well-posedness result does not hold for (S-R-D-C-2) on $BPC^2_{+'}(\mathbb{R})$, if an additional condition is imposed on the initial data, uniform global well-posedness can be established. This follows from qualitative properties of solutions to (S-R-D-C-2) which we now consider.

First, following Chapter 4, let $v : \bar{D}_\infty \to \mathbb{R}$ be the unique global solution to (B-D-C) with $u_0 : \mathbb{R} \to \mathbb{R}$ taken as the initial data for (S-R-D-C-2). It follows from Theorem 4.2 and Remark 4.3 that

$$v(x,t) = \frac{1}{\sqrt{\pi}} \int_{-\infty}^{\infty} u_0(x + 2\sqrt{t}\lambda)e^{-\lambda^2} d\lambda; \quad \forall(x,t) \in \bar{D}_\infty \qquad (9.36)$$

and, moreover, that.

(i) $0 \le v(x,t) \le M_0$; $\quad \forall(x,t) \in \bar{D}_\infty$ (via Theorem 4.7).
(ii) $0 < v(x,t) \le M_0$; $\quad \forall(x,t) \in D_\infty$ (via (9.36) above).
(iii) When

$$u_0(x) \to \begin{cases} l^+ & \text{as } x \to +\infty \\ l^- & \text{as } x \to -\infty, \end{cases}$$

with l^+, $l^- \ge 0$, then

$$v(x,t) \to \begin{cases} l^+ & \text{as } x \to +\infty \\ l^- & \text{as } x \to -\infty \end{cases}$$

uniformly for $t \in [0, T]$, any $T > 0$ (via Remark 4.8).

We can now state;

Proposition‡ 9.20 Let $\bar{u} : \bar{D}_\infty \to \mathbb{R}$ be given by

$$\bar{u}(x,t) = \left(v(x,t)^{(1-p)} + (1-p)t\right)^{\frac{1}{(1-p)}}; \quad \forall(x,t) \in \bar{D}_\infty,$$

where $v : \bar{D}_\infty \to \mathbb{R}$ is as defined in (9.36). Then $\bar{u}$ is a regular supersolution to (S-R-D-C-2) on $\bar{D}_T$ for any $T > 0$.

Proof Since $v : \bar{D}_\infty \to \mathbb{R}$ is non-negative and the solution to (B-D-C), it follows that $\bar{u} : \bar{D}_\infty \to \mathbb{R}$ is continuous and such that $\bar{u}_t$, $\bar{u}_x$ and $\bar{u}_{xx}$ exist and are continuous on D_∞. Furthermore, it follows from (i) above that $\bar{u}$ is bounded on $\bar{D}_T$ for any $T > 0$. Also,

$$\bar{u}(x,0) = u_0(x); \quad \forall x \in \mathbb{R}. \qquad (9.37)$$

Now, for all $(x,t) \in D_\infty$, we have

$$\bar{u}_t(x,t) = \phi^{\frac{p}{1-p}}(x,t) + \frac{\phi^{\frac{p}{(1-p)}}(x,t)v_t(x,t)}{v^p(x,t)},$$

$$\bar{u}_x(x,t) = \frac{\phi^{\frac{p}{(1-p)}}(x,t)v_x(x,t)}{v^p(x,t)},$$

$$\bar{u}_{xx}(x,t) = \frac{\phi^{\frac{p}{(1-p)}}(x,t)v_{xx}(x,t)}{v^p(x,t)} - \frac{p\phi^{\frac{p}{(1-p)}}(x,t)v_x^2(x,t)}{v^{(1+p)}(x,t)}$$

$$+ \frac{p\phi^{\frac{(2p-1)}{(1-p)}}(x,t)v_x^2(x,t)}{v^{2p}(x,t)} \tag{9.38}$$

with

$$\phi(x,t) = v(x,t)^{(1-p)} + (1-p)t; \quad \forall (x,t) \in D_\infty. \tag{9.39}$$

Therefore,

$$N[\bar{u}] \equiv \bar{u}_t - \bar{u}_{xx} - [\bar{u}^p]^+$$

$$= \phi^{\frac{p}{(1-p)}}(x,t) + \frac{\phi^{\frac{p}{(1-p)}}(x,t)}{v^p(x,t)}\left(v_t(x,t) - v_{xx}(x,t)\right)$$

$$+ \frac{pv_x^2(x,t)}{v^{2p}(x,t)}\phi(x,t)^{\frac{(2p-1)}{(1-p)}}\left(v^{(p-1)}(x,t)\phi(x,t) - 1\right)$$

$$- \phi^{\frac{p}{(1-p)}}(x,t)$$

$$= \frac{p(1-p)tv_x^2(x,t)\phi^{\frac{(2p-1)}{(1-p)}}(x,t)}{v^{(1+p)}(x,t)}$$

$$\geq 0 \tag{9.40}$$

for all $(x,t) \in D_\infty$. It then follows from (9.37) and (9.40) that $\bar{u} : \bar{D}_\infty \to \mathbb{R}$ is a regular supersolution to (S-R-D-C-2) on $\bar{D}_T$ for any $T > 0$, as required. $\square$

We next have,

Corollary‡ 9.21 *Let $u : \bar{D}_\infty \to \mathbb{R}$ be the unique global solution to (S-R-D-C-2). Then,*

$$u(x,t) \leq \left(v^{(1-p)}(x,t) + (1-p)t\right)^{\frac{1}{(1-p)}}; \quad \forall (x,t) \in \bar{D}_\infty.$$

Proof The functions $\bar{u}, \underline{u} : \bar{D}_\infty \to \mathbb{R}$ given by

$$\bar{u}(x,t) = \left(v^{(1-p)}(x,t) + (1-p)t\right)^{\frac{1}{(1-p)}},$$

$$\underline{u}(x,t) = u(x,t)$$

for all $(x, t) \in \bar{D}_\infty$, provide a non-negative regular supersolution and a non-negative regular subsolution respectively to (S-R-D-C-2) on $\bar{D}_T$ for any $T > 0$ (via Proposition 9.20). Also, since $\bar{u}(\cdot, 0) = u_0 \in \mathrm{BPC}^2_{+'}(\mathbb{R})$, it then follows from Theorem 9.14 that $\underline{u} \leq \bar{u}$ on $\bar{D}_T$ for any $T > 0$, and so $\underline{u} \leq \bar{u}$ on $\bar{D}_\infty$, as required. □

We can now establish the following property,

Corollary‡ 9.22 *Let $u : \bar{D}_\infty \to \mathbb{R}$ be the unique global solution to (S-R-D-C-2). When $u_0 \in BPC^2_{+'}(\mathbb{R})$ is such that*

$$u_0(x) \to 0 \text{ as } |x| \to \infty,$$

then

$$u(x, t) \to ((1 - p)t)^{\frac{1}{(1-p)}} \text{ as } |x| \to \infty$$

uniformly for $t \in [0, T]$ and any $T > 0$.

Proof This follows directly from Proposition 9.16, Corollary 9.21 and (iii). □

Finally we observe from (9.36), that when $u_0 \in \mathrm{BPC}^2_{+'}(\mathbb{R})$ has compact support, then

$$v(x, t) = O(t^{-1/2}) \text{ as } t \to \infty \tag{9.41}$$

uniformly for all $x \in \mathbb{R}$. We then have,

Corollary‡ 9.23 *Let $u : \bar{D}_\infty \to \mathbb{R}$ be the unique global solution to (S-R-D-C-2). When $u_0 \in BPC^2_{+'}(\mathbb{R})$ has compact support, then*

$$u(x, t) = ((1 - p)t)^{\frac{1}{(1-p)}} + O(t^{\gamma(p)}) \text{ as } t \to \infty,$$

uniformly for $x \in \mathbb{R}$, where

$$\gamma(p) = -\frac{(p^2 - 4p + 1)}{2(1 - p)} \quad \left(< \frac{1}{(1 - p)}\right).$$

Proof It follows from Corollary 9.21 and Proposition 9.16 that

$$((1 - p)t)^{\frac{1}{(1-p)}} \leq u(x, t) \leq \left(v^{(1-p)}(x, t) + (1 - p)t\right)^{\frac{1}{(1-p)}} ; \quad \forall (x, t) \in \bar{D}_\infty. \tag{9.42}$$

With

$$R(x, t) = u(x, t) - ((1 - p)t)^{\frac{1}{(1-p)}},$$

we then have

$$0 \le R(x, t) \le \left(v^{(1-p)}(x, t) + (1 - p)t\right)^{\frac{1}{(1-p)}}$$

$$- ((1 - p)t)^{\frac{1}{(1-p)}} \quad \forall (x, t) \in \bar{D}_\infty. \tag{9.43}$$

Now, from (*i*) and (9.41), there exists a constant $M > 0$ such that

$$0 < v(x, t) < \frac{M}{t^{1/2}}; \quad \forall (x, t) \in D_\infty, \tag{9.44}$$

and so

$$0 \le R(x, t) \le \left(\frac{M^{(1-p)}}{t^{\frac{(1-p)}{2}}} + (1 - p)t\right)^{\frac{1}{(1-p)}} - ((1 - p)t)^{\frac{1}{(1-p)}}; \quad \forall (x, t) \in D_\infty. \tag{9.45}$$

Also, for $t > 0$,

$$\left(\frac{M^{(1-p)}}{t^{\frac{(1-p)}{2}}} + (1 - p)t\right)^{\frac{1}{(1-p)}} = ((1 - p)t)^{\frac{1}{(1-p)}} \left(1 + \frac{M^{(1-p)}}{(1 - p)t^{\frac{(1-p)}{2}+1}}\right)^{\frac{1}{(1-p)}}$$

$$= ((1 - p)t)^{\frac{1}{(1-p)}} + \Psi(t) \tag{9.46}$$

with, via the binomial theorem,

$$\Psi(t) = O(t^{\gamma(p)}) \text{ as } t \to \infty. \tag{9.47}$$

Thus,

$$0 \le R(x, t) \le \Psi(t); \quad \forall (x, t) \in D_\infty,$$

via (9.45) and (9.46), and so

$$0 \le \frac{R(x, t)}{t^{\gamma(p)}} \le \frac{\Psi(t)}{t^{\gamma(p)}}; \quad \forall (x, t) \in D_\infty. \tag{9.48}$$

Now, $\Psi(t)/t^{\gamma(p)}$ is bounded as $t \to \infty$, via (9.47), and so $R(x, t)/t^{\gamma(p)}$ is bounded as $t \to \infty$, uniformly for $x \in \mathbb{R}$. It follows that

$$R(x, t) = O(t^{\gamma(p)}) \text{ as } t \to \infty$$

uniformly for $x \in \mathbb{R}$. Therefore,

$$u(x, t) = ((1 - p)t)^{\frac{1}{(1-p)}} + O(t^{\gamma(p)}) \text{ as } t \to \infty$$

uniformly for $x \in \mathbb{R}$, as required. $\qquad\square$

A concluding statement can now be made concerning uniform global well-posedness and stability of a restricted version of (S-R-D-C-2), namely,

Remark 9.24 Consider the set $\mathcal{A}$ given by

$$\mathcal{A} = \left\{ u_0 \in \mathrm{BPC}^2_{+'}(\mathbb{R}) : u_0 \text{ has compact support} \right\}.$$

Then (S-R-D-C-2) is uniformly globally well-posed on $\mathcal{A}$ when $0 < p < 2 - \sqrt{3}$, which follows from Corollary 9.18 and Corollary 9.23, noting that $\gamma(p) < 0$ when $0 < p < 2 - \sqrt{3}$. Moreover, under these conditions the global solution is asymptotically stable with respect to perturbations in the initial data $\delta u_0 \in \mathcal{A}$. Note that $0 < p < 2 - \sqrt{3}$ is a sufficient condition for Remark 9.24 to hold. ⌐

9.3 $f(u) = [u^p]^+[(1-u)^q]^+$

In this section we consider the (B-R-D-C) problem when the reaction function $f : \mathbb{R} \to \mathbb{R}$ is given by

$$f(u) = [u^p]^+[(1-u)^q]^+ = \begin{cases} u^p(1-u)^q & ; u \in [0,1] \\ 0 & ; u \notin [0,1], \end{cases} \tag{9.49}$$

where $p, q \in (0,1)$. In particular, we restrict attention to initial data $u_0 \in \mathrm{BPC}^2_{+'}(\mathbb{R})$, with

$$\sup_{x \in \mathbb{R}} \{u_0(x)\} = M_0 > 0, \quad \inf_{x \in \mathbb{R}} \{u_0(x)\} = m_0 \geq 0. \tag{9.50}$$

The aim of this section is to provide a uniform global well-posedness result for this specific (B-R-D-C). This problem is similar to (S-R-D-C-2) in that the technical condition (9.50) is necessary. Throughout the section, this (B-R-D-C) will be referred to as (S-R-D-C-3). It should be noted that the problem (S-R-D-C-3) has been studied in [36], and briefly reviewed in [75], in the Lipschitz case when $p, q \geq 1$ and in [33] when $0 < p < 1$ and $q = 1$. Aspects of this subsection can also be found in [50]. For convenience, we define $\gamma = \frac{p}{p+q}$ and observe that

$$\sup_{u \in \mathbb{R}} f(u) = f(\gamma) = (\gamma)^p (1-\gamma)^q, \tag{9.51}$$

and that $f : (-\infty, \gamma] \to \mathbb{R}$ is non-decreasing and $f : [\gamma, \infty) \to \mathbb{R}$ is non-increasing.

Before we proceed with the uniform global well-posedness result for (S-R-D-C-3), we highlight several important features of the reaction function (9.49).

Proposition 9.25 *The function $f : \mathbb{R} \to \mathbb{R}$ given by (9.49) satisfies $f \in H_\alpha$ where $\alpha = min\{p, q\}$. Moreover, on any closed bounded interval $E \subset \mathbb{R}$, the Hölder constant satsfies $k_H = 1$.*

Proof First consider the closed bounded interval $[0, \gamma]$. Observe that $f : \mathbb{R} \to \mathbb{R}$ is differentiable on $(0, \gamma)$, with a derivative which satisfies

$$\frac{df(u)}{du} = pu^{p-1}(1-u)^q - qu^p(1-u)^{q-1} \le pu^{p-1} = \frac{du^p}{du}, \quad (9.52)$$

for all $u \in (0, \gamma)$. Therefore, for all $x, y \in [0, \gamma]$ such that $x \le y$, an integration of (9.52) from x to y yields

$$|f(y) - f(x)| = f(y) - f(x) = \int_x^y \frac{df(u)}{du} du \le \int_x^y \frac{du^p}{du} du$$
$$= y^p - x^p \le (y - x)^p \le (y - x)^\alpha = |y - x|^\alpha \quad (9.53)$$

Next consider the interval $[\gamma, 1]$. Observe that $f : \mathbb{R} \to \mathbb{R}$ is differentiable on $(\gamma, 1)$, with a derivative that satisfies

$$\frac{df(u)}{du} = pu^{p-1}(1-u)^q - qu^p(1-u)^{q-1} \ge -q(1-u)^{q-1} = \frac{d(1-u)^q}{du}$$
$$(9.54)$$

for all $u \in (\gamma, 1)$. Again, for all $x, y \in [\gamma, 1]$ such that $x \le y$, an integration of (9.54) from x to y yields

$$|f(y) - f(x)| = f(x) - f(y) = -\int_x^y \frac{df(u)}{du} du \le -\int_x^y \frac{d(1-u)^q}{du} du$$
$$= (1-x)^q - (1-y)^q$$
$$\le (y-x)^q = |y-x|^q \le |y-x|^\alpha. \quad (9.55)$$

Now, for every $u \in (0, \gamma)$ there exists $\tilde{u} \in (\gamma, 1)$ such that $f(u) = f(\tilde{u})$. Therefore, for all $x, y \in [0, 1]$ such that $0 \le x \le \gamma \le y \le 1$, with $0 \le \tilde{y} \le \gamma \le \tilde{x} \le 1$, then (9.55) and (9.53) imply

$$|f(y) - f(x)| = |f(\tilde{y}) - f(x)|$$
$$= |f(y) - f(\tilde{x})| \le \min\{|y - \tilde{x}|^\alpha, |\tilde{y} - x|^\alpha\} \le |y - x|^\alpha.$$
$$(9.56)$$

Finally, since $f(u) = 0$ for all $u \notin (0, 1)$, inequalities (9.53), (9.55) and (9.56) ensure that

$$|f(y) - f(x)| \le |y - x|^\alpha$$

for any $x, y \in \mathbb{R}$, as required. $\qquad\square$

Proposition 9.26 *For any $x, y \in \mathbb{R}$, such that $x \le y$, $f : \mathbb{R} \to \mathbb{R}$ given by (9.49) satisfies*

$$f(y) - f(x) \le (y - x)^p.$$

Proof First, observe that $f : \mathbb{R} \to \mathbb{R}$ is differentiable on $(0, 1)$ and

$$\frac{df(s)}{ds} = ps^{p-1}(1-s)^q - qs^p(1-s)^{q-1} \le ps^{p-1} \le \frac{d(s^p)}{ds}; \quad \forall s \in (0,1).$$
(9.57)

An integration of (9.57) with respect to s from x to y, where $0 \le x \le y \le 1$, yields

$$f(y) - f(x) \le y^p - x^p \le (y - x)^p.$$
(9.58)

Furthermore, for any $x, y \notin (0, 1)$, we have

$$f(y) - f(x) = 0 \le (y - x)^p.$$
(9.59)

The result follows from (9.58) and (9.59). □

Proposition 9.27 *For any $x, y \in [0, \infty)$, with $y > x$, then $f : \mathbb{R} \to \mathbb{R}$ given by (9.49) satisfies*

$$f(y) - f(x) \le p\theta^{p-1}(y - x)$$

for some $\theta \in (x, y)$.

Proof Observe that $f : \mathbb{R} \to \mathbb{R}$ given by (9.49) is differentiable in $(0, 1)$, and, via (9.57),

$$f'(s) \le ps^{p-1}; \quad \forall s \in (0, 1).$$
(9.60)

Now, for $x, y \in [0, 1]$ with $x < y$, the mean value theorem and (9.60) establish that there exists $\theta \in (x, y)$ such that

$$f(y) - f(x) = f'(\theta)(y - x) \le p\theta^{p-1}(y - x).$$
(9.61)

The result follows on noting that $f(s) = 0$ for all $s \in [1, \infty)$. □

We now proceed to existence and uniqueness results for (S-R-D-C-3). We first define the function $f_\eta : \mathbb{R} \to \mathbb{R}$, for any $\eta \in (0, \gamma]$ such that

$$f_\eta(u) = \begin{cases} f(\eta) & ; u < \eta \\ f(u) & ; u \ge \eta, \end{cases}$$
(9.62)

where $f : \mathbb{R} \to \mathbb{R}$ is given by (9.49). We have the following;

Proposition 9.28 *Let $f : \mathbb{R} \to \mathbb{R}$ and $f_\eta : \mathbb{R} \to \mathbb{R}$ be given by (9.49) and (9.62) respectively. Then $f_\eta \in L_u$ and*

$$f_\eta(u) \ge f(u); \quad \forall u \in \mathbb{R}.$$
(9.63)

Proof Observe that for $x, y \in \mathbb{R}$, with $\eta \leq x \leq y \leq 1$, via (9.62) and the mean value theorem, there exists $\theta \in (x, y)$, such that

$$f_\eta(y) - f_\eta(x) = f'(\theta)(y - x) \leq f'(\eta)(y - x). \tag{9.64}$$

Since $f_\eta(s) = 0$ for all $s \in [1, \infty)$ and $f_\eta(s) = f(\eta)$ for all $s \in (-\infty, \eta]$, then (9.64) holds for any $x, y \in \mathbb{R}$ with $x \leq y$. Thus $f_\eta \in L_u$. Moreover, via (9.49), $f : \mathbb{R} \to \mathbb{R}$ is non-decreasing on $(-\infty, \eta]$, and (9.63) follows. $\square$

We now establish a priori bounds for (S-R-D-C-3).

Proposition 9.29 *Let $u : \bar{D}_T \to \mathbb{R}$ be any solution to (S-R-D-C-3). Then,*

$$m_0 \leq u(x, t) \leq \max\{M_0, 1\}, \quad \forall (x, t) \in \bar{D}_T.$$

Proof Since $f : \mathbb{R} \to \mathbb{R}$ given by (9.49) is non-negative, then the left inequality is obtained via Theorem 3.6. Now define $\bar{u}, \underline{u} : \bar{D}_T \to \mathbb{R}$ to be

$$\bar{u}(x, t) = \max\{M_0, 1\}, \quad \underline{u}(x, t) = u(x, t); \quad \forall (x, t) \in \bar{D}_T. \tag{9.65}$$

It follows from Proposition 9.28 that

$$\left. \begin{array}{r} \bar{u}_t - \bar{u}_{xx} - f_\eta(\bar{u}) \geq 0 \\ \underline{u}_t - \underline{u}_{xx} - f_\eta(\underline{u}) = f(\underline{u}) - f_\eta(\underline{u}) \leq 0 \end{array} \right\} \quad \forall (x, t) \in D_T \tag{9.66}$$

$$\underline{u}(x, 0) \leq \bar{u}(x, 0); \quad \forall x \in \mathbb{R}. \tag{9.67}$$

Therefore $\bar{u}$ and $\underline{u}$ can be taken as a regular supersolution and a regular subsolution respectively for (B-R-D-C) with reaction function $f_\eta : \mathbb{R} \to \mathbb{R}$ and it then follows from Theorem 7.1 and (9.65) that

$$u(x, t) \leq \max\{M_0, 1\}; \quad \forall (x, t) \in \bar{D}_T,$$

as required. $\square$

Remark 9.30 It follows from Proposition 9.29 that (S-R-D-C-3) is uniformly a priori bounded on $\bar{D}_T$ for any $T > 0$ and hence is a priori bounded on $\bar{D}_\infty$.

We can now state,

Theorem$_\ddagger$ 9.31 *(S-R-D-C-3) has a global constructed maximal solution $\bar{u}^c : \bar{D}_\infty \to \mathbb{R}$ and a global constructed minimal solution $\underline{u}^c : \bar{D}_\infty \to \mathbb{R}$ for any $u_0 \in BPC_{+'}^2(\mathbb{R})$. Moreover,*

$$m_0 \leq \underline{u}^c(x, t) \leq \bar{u}^c(x, t) \leq \max\{M_0, 1\}; \quad \forall (x, t) \in \bar{D}_\infty.$$

Proof Existence follows from Remark 9.30, Proposition 9.25 and Theorem 8.25. The bounds follow from Proposition 9.29. □

Before we can establish a uniqueness argument, we require an improved lower bound for (S-R-D-C-3), similar to Corollary 9.13 for (S-R-D-C-2).

Theorem‡ 9.32 *The constructed minimal solution* $\underline{u}^c : \bar{D}_\infty \to \mathbb{R}$ *to (S-R-D-C-3) satisfies*

$$\underline{u}^c(x,t) \geq ((1-p)ct)^{1/(1-p)}, \quad \forall (x,t) \in \bar{D}_{T_c},$$

where $T_c = \frac{(1-(1/2)^{(1-p)})(1-c^{1/q})^{(1-p)}}{(1-p)}$ *for any* $c \in (0,1)$.

Proof To begin, fix $c \in (0,1)$ and let $\bar{u}^c, \underline{u}^c : \bar{D}_\infty \to \mathbb{R}$ be the constructed maximal and constructed minimal solution respectively to (S-R-D-C-3) with initial data $u_0 \in \mathrm{BPC}^2_{+'}(\mathbb{R})$, as in Theorem 9.31. Now consider the (B-R-D-C) problem with reaction function $\hat{f} : \mathbb{R} \to \mathbb{R}$ given by

$$\hat{f}(u) = \begin{cases} cu^p & ; u \in [0, u_c] \\ f(u) & ; u \notin [0, u_c] \end{cases} \leq f(u); \quad \forall u \in \mathbb{R}, \qquad (9.68)$$

with $f : \mathbb{R} \to \mathbb{R}$ given by (9.49), $u_c = (1 - c^{1/q}) \in (0,1)$, and initial data $\hat{u}_0 \in \mathrm{BPC}^2_{+'}(\mathbb{R})$ given by

$$\hat{u}_0(x) = \frac{u_c u_0(x)}{2\max\{1, M_0\}} \leq \min\left\{\tfrac{1}{2}u_c, u_0(x)\right\}; \quad \forall x \in \mathbb{R}. \qquad (9.69)$$

It follows from Proposition 9.25 and (9.68) that $\hat{f} \in H_\alpha$, where $\alpha = \min\{p, q\}$. Now let $u : \bar{D}_T \to \mathbb{R}$ be any solution to (B-R-D-C) with $\hat{f}$ and $\hat{u}_0$, then since $\hat{f} : \mathbb{R} \to \mathbb{R}$ is non-negative, via Theorem 3.6, we have

$$u(x,t) \geq 0 \quad \forall (x,t) \in \bar{D}_T. \qquad (9.70)$$

Moreover, since $u : \bar{D}_T \to \mathbb{R}$ is a solution to (B-R-D-C) with $\hat{f}$ and $\hat{u}_0$, via (9.68), it follows that

$$u_t - u_{xx} - f(u) = \hat{f}(u) - f(u) \leq 0, \quad \forall (x,t) \in D_T. \qquad (9.71)$$

It follows from (9.69) and (9.71) that $u : \bar{D}_T \to \mathbb{R}$ is a R-S-B to (S-R-D-C-3) with initial data $u_0 \in \mathrm{BPC}^2_{+'}(\mathbb{R})$. Therefore, via Proposition 8.26,

$$u(x,t) \leq \bar{u}^c(x,t); \quad \forall (x,t) \in \bar{D}_T. \qquad (9.72)$$

Therefore, via (9.72) and Theorem 9.31, we have

$$u(x,t) \leq \max\{1, M_0\}; \quad \forall (x,t) \in \bar{D}_T, \qquad (9.73)$$

and so, from (9.70) and (9.73), we conclude that (B-R-D-C) is a priori bounded on $\bar{D}_T$ uniformly in $T > 0$. Thus it follows from Theorem 8.25 that there exists a constructed minimal solution $\hat{\underline{u}} : \bar{D}_\infty \to \mathbb{R}$ to (B-R-D-C). Now, since $\hat{f} \in H_\alpha$, whilst $\hat{\underline{u}} : \bar{D}_\infty \to \mathbb{R}$ is the constructed minimal solution to (B-R-D-C) and $\underline{u}^c : \bar{D}_\infty \to \mathbb{R}$ is a R-S-P to (B-R-D-C) with initial data $\hat{u}_0 \in \text{BPC}^2_{+'}(\mathbb{R})$, then, via Proposition 8.26, we have

$$\hat{\underline{u}}(x,t) \le \underline{u}^c(x,t); \quad \forall (x,t) \in \bar{D}_\infty. \tag{9.74}$$

Next, since $\hat{\underline{u}} : \bar{D}_T \to \mathbb{R}$ is a solution to (B-R-D-C) on $\bar{D}_T$, then via (9.68), we have

$$\hat{\underline{u}}_t - \hat{\underline{u}}_{xx} - \hat{\underline{u}}^p = \hat{f}(\hat{\underline{u}}) - \hat{\underline{u}}^p \le 0; \quad \forall (x,t) \in D_T. \tag{9.75}$$

It follows from (9.75) and (9.69) that $\hat{\underline{u}}$ is a R-S-B to (S-R-D-C-2) with initial data $\hat{u}_0 \in \text{BPC}^2_{+'}(\mathbb{R})$. Thus, via Theorem 9.14, we have

$$0 \le \hat{\underline{u}}(x,t) \le u_2(x,t); \quad \forall (x,t) \in \bar{D}_T,$$

where $u_2 : \bar{D}_T \to \mathbb{R}$ is the unique solution to (S-R-D-C-2) with initial data $\hat{\underline{u}}_0 \in \text{BPC}^2_{+'}(\mathbb{R})$. It then follows from (9.69) and Theorem 9.16 that

$$0 \le \hat{\underline{u}}(x,t) \le \left((1-p)t + \left(\frac{u_c}{2} \right)^{(1-p)} \right)^{1/(1-p)}; \quad \forall (x,t) \in \bar{D}_T, \tag{9.76}$$

and so

$$0 \le \hat{\underline{u}}(x,t) \le u_c; \quad \forall (x,t) \in \bar{D}_{T_c}, \tag{9.77}$$

where $T_c = \frac{(1-(1/2)^{(1-p)})(1-c^{1/q})^{(1-p)}}{(1-p)}$. It now follows from (9.68), (9.69) and (9.77) that $\hat{\underline{u}} : \bar{D}_{T_c} \to \mathbb{R}$ is a solution to the (B-R-D-C) problem with reaction function $\check{f} : \mathbb{R} \to \mathbb{R}$ given by

$$\check{f}(u) = c[u^p]^+; \quad \forall u \in \mathbb{R}, \tag{9.78}$$

with initial data $\hat{\underline{u}}_0 \in \text{BPC}^2_{+'}(\mathbb{R})$. Next, define the function $z : \bar{D}_{cT_c} \to \mathbb{R}$ to be

$$z(\tilde{x}, \tilde{t}) = \hat{\underline{u}}(x,t); \quad \forall (\tilde{x}, \tilde{t}) \in \bar{D}_{cT_c}, \tag{9.79}$$

where $\tilde{x} = c^{1/2} x$ and $\tilde{t} = ct$. We observe from (9.78) and (9.79) that

$$z_{\tilde{t}} - z_{\tilde{x}\tilde{x}} - [z^p]^+ = 0; \quad \forall (\tilde{x}, \tilde{t}) \in D_{cT_c}, \tag{9.80}$$

with initial data $z(\cdot, 0) \in \text{BPC}^2_{+'}(\mathbb{R})$. It follows from (9.80) that $z : \bar{D}_{cT_c} \to \mathbb{R}$ is the unique solution to (S-R-D-C-2) on $\bar{D}_{cT_c}$ with initial data $z(\cdot, 0) \in \text{BPC}^2_{+'}(\mathbb{R})$. Therefore, via Theorem 9.16, we have

$$z(\tilde{x}, \tilde{t}) \ge ((1-p)\tilde{t})^{1/(1-p)}; \quad \forall (\tilde{x}, \tilde{t}) \in \bar{D}_{cT_c},$$

and hence, via (9.79), it follows that

$$\hat{u}(x, t) \geq ((1-p)ct)^{1/(1-p)}; \quad \forall (x, t) \in \bar{D}_{T_c}. \tag{9.81}$$

The result follows from (9.81) and (9.74).   □

We can now establish a uniqueness result for (S-R-D-C-3). The proof follows a similar approach to that of Aguirre and Escobedo in [5].

Theorem‡ 9.33 (Uniqueness) *The constructed minimal solution $\underline{u}^c : \bar{D}_\infty \to \mathbb{R}$ to (S-R-D-C-3), is the unique solution to (S-R-D-C-3).*

Proof For $u_0 \in \text{BPC}^2_{+\prime}(\mathbb{R})$ with $m_0 > 0$ in (9.50), via Theorem 9.31, any corresponding solution $u : \bar{D}_T \to \mathbb{R}$ to (S-R-D-C-3) is a solution to (B-R-D-C) with reaction function $f_\eta : \mathbb{R} \to \mathbb{R}$ with $\eta = \min\{m_0, \gamma\}$. It follows from Proposition 9.28 and Theorem 7.2 that $u = \underline{u}^c$ on $\bar{D}_T$. Thus $\underline{u}^c : \bar{D}_\infty \to \mathbb{R}$ is the unique solution to (S-R-D-C-3) in this case.

Now, consider $u_0 \in \text{BPC}^2_{+\prime}(\mathbb{R})$ with $m_0 = 0$ in (9.50). Suppose that $\bar{u}^c : \bar{D}_\infty \to \mathbb{R}$ and $\underline{u}^c : \bar{D}_\infty \to \mathbb{R}$ are as in Theorem 9.31. Then, via Proposition 9.26 and the Hölder Equivalence Lemma 5.10, we have

$$(\bar{u}^c - \underline{u}^c)(x, t) = \frac{1}{\sqrt{\pi}} \int_0^t \int_{-\infty}^\infty (f(\bar{u}^c) - f(\underline{u}^c)) \left(x + 2\sqrt{t-\tau}\lambda, \tau\right) e^{-\lambda^2} d\lambda d\tau \tag{9.82}$$

$$\leq \frac{1}{\sqrt{\pi}} \int_0^t \int_{-\infty}^\infty (\bar{u}^c - \underline{u}^c)^p \left(x + 2\sqrt{t-\tau}\lambda, \tau\right) e^{-\lambda^2} d\lambda d\tau$$

$$\leq \frac{1}{\sqrt{\pi}} \int_0^t \int_{-\infty}^\infty ||(\bar{u}^c - \underline{u}^c)(\cdot, \tau)||_B^p e^{-\lambda^2} d\lambda d\tau$$

$$\leq \int_0^t ||(\bar{u}^c - \underline{u}^c)(\cdot, \tau)||_B^p d\tau \tag{9.83}$$

for all $(x, t) \in \bar{D}_T$ and any $T > 0$, on noting, via Corollary 5.16, that $\underline{u}^c, \bar{u}^c : \bar{D}_\infty \to \mathbb{R}$ are uniformly continuous on $\bar{D}_T$, and so $||(\bar{u}^c - \underline{u}^c)(\cdot, t)||_B$ is continuous for $t \in [0, T]$. Moreover, the right hand side of (9.83) is independent of x, from which we obtain

$$||(\bar{u}^c - \underline{u}^c)(\cdot, t)||_B \leq \int_0^t ||(\bar{u}^c - \underline{u}^c)(\cdot, \tau)||_B^p d\tau; \quad \forall t \in [0, T],$$

which gives, after an integration,

$$||(\bar{u}^c - \underline{u}^c)(\cdot, t)||_B \leq ((1-p)t)^{1/(1-p)}; \quad \forall t \in [0, T]. \tag{9.84}$$

Now, via Proposition 9.27, Theorem 9.32 and the mean value theorem, for any $(s, \tau) \in \bar{D}_{T^*}$ there exists $\theta \in [\underline{u}^c(s, \tau), \overline{u}^c(s, \tau)]$ such that

$$f(\overline{u}^c(s, \tau)) - f(\underline{u}^c(s, \tau)) = f'(\theta)(\overline{u}^c(s, \tau) - \underline{u}^c(s, \tau)) \tag{9.85}$$

$$\leq p\theta^{p-1}(\overline{u}^c(s, \tau) - \underline{u}^c(s, \tau))$$

$$\leq p((1-p)c\tau)^{-1}(\overline{u}^c(s, \tau) - \underline{u}^c(s, \tau))$$

$$\leq \frac{p}{(1-p)c\tau}||(\overline{u}^c - \underline{u}^c)(\cdot, \tau)||_B, \tag{9.86}$$

where $T^* = T_c$ is defined by Theorem 9.32 for $c \in (0, 1)$, with c chosen so that

$$0 < p < c < 1. \tag{9.87}$$

On substituting (9.86) into (9.82), we have, for $(x, t) \in D_{T^*}$,

$$(\overline{u}^c - \underline{u}^c)(x, t)$$

$$\leq \frac{1}{\sqrt{\pi}} \int_0^t \int_{-\infty}^{\infty} \frac{p}{(1-p)c\tau} ||(\overline{u}^c - \underline{u}^c)(\cdot, \tau)||_B e^{-\lambda^2} d\lambda d\tau; \quad \forall (x, t) \in D_{T^*}$$

and so

$$||(\overline{u}^c - \underline{u}^c)(\cdot, t)||_B \leq \frac{p}{(1-p)c} \int_0^t \tau^{-1} ||(\overline{u}^c - \underline{u}^c)(\cdot, \tau)||_B d\tau; \quad \forall t \in [0, T^*], \tag{9.88}$$

on noting that the right hand side of (9.88) is integrable via (9.84) and Lemma 5.5, and the limit of the right hand side implied at $t = 0$. Next we define the function $w : [0, T^*] \to \mathbb{R}$ to be

$$w(t) = \begin{cases} \int_0^t \tau^{-1} ||(\overline{u}^c - \underline{u}^c)(\cdot, \tau)||_B d\tau & ; t \in (0, T^*] \\ 0 & ; t = 0. \end{cases} \tag{9.89}$$

We note that w is non-negative, continuous and differentiable (via Corollary 5.16). The inequality (9.88) can be re-written as

$$w'(s) - \frac{p}{c(1-p)s} w(s) \leq 0; \quad \forall s \in (0, T^*]. \tag{9.90}$$

This may be re-written as

$$\left(w(s)s^{-\frac{p}{c(1-p)}}\right)' \leq 0; \quad \forall s \in (0, T^*]. \tag{9.91}$$

We now integrate (9.91) from $s = \epsilon$ to $s = t$ (with $0 < \epsilon < t \leq T^*$) to obtain

$$w(t) \leq w(\epsilon) \left(\frac{t}{\epsilon}\right)^{\frac{p}{c(1-p)}}; \quad \forall 0 < \epsilon < t \leq T^*. \tag{9.92}$$

Next we substitute the bound in (9.84) into (9.89), which gives

$$w(\epsilon) = \int_0^\epsilon \tau^{-1}||(\overline{u}^c - \underline{u}^c)(\cdot, \tau)||_B d\tau$$
$$\leq \int_0^\epsilon (1-p)^{1/(1-p)} \tau^{1/(1-p)-1} d\tau$$
$$= (1-p)^{(2-p)/(1-p)} \epsilon^{1/(1-p)} \tag{9.93}$$

for $0 < \epsilon < t \leq T^*$. Finally, upon substituting (9.93) into (9.92), we obtain

$$w(t) \leq (1-p)^{(2-p)/(1-p)} T^{*p/c(1-p)} \epsilon^{\frac{1}{(1-p)}(1-\frac{p}{c})}; \quad \forall \, 0 < \epsilon < t \leq T^*. \tag{9.94}$$

Now, via (9.87), upon letting $\epsilon \to 0$ in (9.94), we obtain

$$w(t) = 0; \quad \forall t \in [0, T^*]. \tag{9.95}$$

Therefore, via (9.95), (9.89) and (9.88), we have

$$||(\overline{u}^c - \underline{u}^c)(\cdot, t)||_B = 0; \quad \forall t \in [0, T^*],$$

and hence

$$\overline{u}^c(x, t) = \underline{u}^c(x, t); \quad \forall (x, t) \in \bar{D}_{T^*}. \tag{9.96}$$

When $T \leq T^*$, the proof is complete, so now let $T > T^*$. All that remains is to establish the uniqueness of $\underline{u}^c$ on $\bar{D}_T^{T^*}$. To this end, consider the functions $\underline{u}_{T^*}^c, \overline{u}_{T^*}^c : \bar{D}_{T-T^*} \to \mathbb{R}$ defined to be

$$\left.\begin{array}{l} \underline{u}_{T^*}^c(x, t) = \underline{u}^c(x, t + T^*) \\ \overline{u}_{T^*}^c(x, t) = \overline{u}^c(x, t + T^*) \end{array}\right\}; \quad \forall (x, t) \in \bar{D}_{T-T^*}. \tag{9.97}$$

Following from the definition of $\overline{u}_{T^*}^c$ and $\underline{u}_{T^*}^c$, Theorem 9.32 and (9.96), we have, for $c \in (0, 1)$ as in (9.87),

$$0 < ((1-p)cT^*)^{1/(1-p)} \leq \underline{u}_{T^*}^c(x, 0) = \overline{u}_{T^*}^c(x, 0); \quad \forall x \in \mathbb{R}, \tag{9.98}$$

where $\underline{u}_{T^*}^c(\cdot, 0), \overline{u}_{T^*}^c(\cdot, 0) \in \text{BPC}_{+'}^2(\mathbb{R})$, via Proposition 9.25, Theorem 9.31, Lemma 5.12 and Lemma 5.15. Moreover, from Theorem 9.31 and (9.97), it follows that

$$\underline{u}_{T^*}^c(x, t) \leq \overline{u}_{T^*}^c(x, t); \quad \forall (x, t) \in \bar{D}_{T-T^*}. \tag{9.99}$$

Additionally, both $\underline{u}_{T^*}^c$ and $\overline{u}_{T^*}^c$ are bounded, twice continuously differentiable with respect to x and once with respect to t on $\bar{D}_{T-T^*}$. Now, since $\underline{u}_{T^*}^c$ satisfies

$$\underline{u}_{T^*\,t}^c - \underline{u}_{T^*\,xx}^c \geq f(\underline{u}_{T^*}^c) \geq 0; \quad \forall (x, t) \in D_{T-T^*},$$

Theorem 3.6, in conjunction with (9.98) and (9.99) establishes that

$$0 < ((1-p)cT^*)^{1/(1-p)} \le \underline{u}_{T^*}^c(x,t) \le \overline{u}_{T^*}^c(x,t); \quad \forall(x,t) \in \bar{D}_{T-T^*}.$$
$$(9.100)$$

Observe that since $\overline{u}_{T^*}^c$ and $\underline{u}_{T^*}^c$ solve (S-R-D-C-3), then via (9.100) and Proposition 9.28,

$$\left.\begin{aligned} \underline{u}_{T^*t}^c - \underline{u}_{T^*xx}^c - f_\eta(\underline{u}_{T^*}^c) \ge 0 \\ \overline{u}_{T^*t}^c - \overline{u}_{T^*xx}^c - f_\eta(\overline{u}_{T^*}^c) \le 0 \end{aligned}\right\}; \quad \forall(x,t) \in D_{T-T^*}, \qquad (9.101)$$

where $f_\eta : \mathbb{R} \to \mathbb{R}$ is defined as in Proposition 9.28, with η chosen as

$$\eta = \min\{((1-p)cT^*)^{1/(1-p)}, \gamma\}.$$

Finally, via Proposition 9.28, $f_\eta \in L_u$, and also, via (9.101) and (9.98), $\underline{u}_{T^*}^c : \bar{D}_{T-T^*} \to \mathbb{R}$ and $\overline{u}_{T^*}^c : \bar{D}_{T-T^*} \to \mathbb{R}$ are a R-S-P and a R-S-B to (B-R-D-C) with reaction function f_η and $u_0 = \overline{u}^c(\cdot, T^*) = \underline{u}^c(\cdot, T^*) \in \mathrm{BPC}_+^2(\mathbb{R})$. It follows from Theorem 7.1 that

$$\underline{u}_{T^*}^c(x,t) \ge \overline{u}_{T^*}^c(x,t); \quad \forall(x,t) \in \bar{D}_{T-T^*}. \qquad (9.102)$$

It then follows from (9.99) and (9.102) that

$$\underline{u}_{T^*}^c(x,t) = \overline{u}_{T^*}^c(x,t); \quad \forall(x,t) \in \bar{D}_{T-T^*}. \qquad (9.103)$$

Finally, equations (9.103), (9.97) and (9.96) give

$$\overline{u}^c(x,t) = \underline{u}^c(x,t); \quad \forall(x,t) \in \bar{D}_T.$$

This holds for any $T > 0$, and so

$$\overline{u}^c(x,t) = \underline{u}^c(x,t); \quad \forall(x,t) \in \bar{D}_\infty,$$

as required. $\qquad\qquad\qquad\qquad\qquad\qquad\qquad\qquad\qquad\qquad\qquad\qquad \Box$

We now have the following useful comparison theorem for (S-R-D-C-3).

Corollary$_\ddagger$ 9.34 *Let $\overline{u}, \underline{u} : \bar{D}_T \to \mathbb{R}$ be a R-S-P and a R-S-B to (S-R-D-C-3). Then $\underline{u}(x,t) \le \overline{u}(x,t)$ for all $(x,t) \in \bar{D}_T$.*

Proof Follows from Proposition 9.25, Theorem 8.26 and Theorem 9.33. $\qquad \Box$

At this stage we can now consider continuous dependence for (S-R-D-C-3).

Theorem‡ **9.35** (Global Continuous Dependence) *Given* $\epsilon > 0$, $T \in (0, \infty)$ *and* $u_{10} \in BPC^2_{+\prime}(\mathbb{R})$, *there exists* $\delta > 0$, *such that for any* $u_{20} \in BPC^2_{+\prime}(\mathbb{R})$ *which satisfies*

$$\|u_{20} - u_{10}\|_B < \delta,$$

the corresponding unique solutions $u_1, u_2 : \bar{D}_T \to \mathbb{R}$ *to (S-R-D-C-3) are such that*

$$\|u_2 - u_1\|_A < \epsilon.$$

Proof Consider $u_{30} \in BPC^2_{+\prime}(\mathbb{R})$, given by

$$u_{30}(x) = u_{10}(x) + \frac{1}{2}\delta; \quad \forall x \in \mathbb{R}, \tag{9.104}$$

with $\delta > 0$. It follows from Theorem 9.31 and Theorem 9.33 that there exists $u_3 : \bar{D}_T \to \mathbb{R}$ that uniquely solves (S-R-D-C-3) with initial data $u_{30} \in BPC^2_{+\prime}(\mathbb{R})$. Now, for any $u_{20} \in BPC^2_{+\prime}(\mathbb{R})$ such that $\|u_{20} - u_{10}\|_B < \frac{\delta}{2}$, then

$$0 < u_{30}(x) - u_{i0}(x) < \delta; \quad \forall x \in \mathbb{R}, \tag{9.105}$$

with $i = 1, 2$. It then follows from taking $u_3 : \bar{D}_T \to \mathbb{R}$ as a R-S-P and $u_i : \bar{D}_T \to \mathbb{R}$ $(i = 1, 2)$ as a R-S-B in Corollary 9.34, that

$$\max\{u_1(x, t), u_2(x, t)\} \le u_3(x, t); \quad \forall (x, t) \in \bar{D}_T. \tag{9.106}$$

Now, via the Hölder Equivalence Lemma 5.10, (9.105), (9.106) and Proposition 9.26 for $i = 1, 2$,

$$0 \le (u_3 - u_i)(x, t)$$

$$\le \delta + \frac{1}{\sqrt{\pi}} \int_0^t \int_{-\infty}^\infty (f(u_3) - f(u_i)) \left(x + 2\sqrt{t - \tau}\lambda, \tau\right) e^{-\lambda^2} d\lambda d\tau$$

$$\le \delta + \frac{1}{\sqrt{\pi}} \int_0^t \int_{-\infty}^\infty (u_3 - u_i)^p \left(x + 2\sqrt{t - \tau}\lambda, \tau\right) e^{-\lambda^2} d\lambda d\tau$$

$$\le \delta + \frac{1}{\sqrt{\pi}} \int_0^t \int_{-\infty}^\infty \|(u_3 - u_i)(\cdot, \tau)\|_B^p e^{-\lambda^2} d\lambda d\tau$$

$$\le \delta + \int_0^t \|(u_3 - u_i)(\cdot, \tau)\|_B^p d\tau \tag{9.107}$$

for all $(x, t) \in \bar{D}_T$. Therefore, since the right hand side of (9.107) is independent of x, we have

$$\|(u_3 - u_i)(\cdot, t)\|_B \le \delta + \int_0^t \|(u_3 - u_i)(\cdot, \tau)\|_B^p d\tau; \quad \forall t \in [0, T], \tag{9.108}$$

from which we obtain (noting that $\|(u_3 - u_i)(\cdot, t)\|_B$ is continuous for $t \in [0, T]$ via Corollary 5.16),

$$\|(u_3 - u_i)(\cdot, t)\|_B \le (\delta^{(1-p)} + (1-p)t)^{1/(1-p)}, \quad (i = 1, 2); \quad \forall t \in [0, T]. \tag{9.109}$$

Now take δ sufficiently small so that $T_\delta = \frac{\delta^{(1-p)}}{(1-p)} < T$ and it follows from (9.109) that

$$\|(u_3 - u_i)(\cdot, t)\|_B \le (\delta^{(1-p)} + \delta^{(1-p)})^{1/(1-p)}$$
$$\le 2^{1/(1-p)}\delta, \quad (i = 1, 2); \quad \forall t \in [0, T_\delta]. \tag{9.110}$$

Next, fix $c \in (0, 1)$ such that $p < c < 1$, and it follows, via Theorem 9.32, that there exists $T_c > 0$ (as given in Theorem 9.32) which is *independent of* δ, such that

$$u_i(x, t) \ge ((1-p)ct)^{1/(1-p)}, \quad (i = 1, 2, 3); \quad \forall (x, t) \in \bar{D}_{T_c}. \tag{9.111}$$

Now take δ sufficiently small so that $T_\delta < T_c$, and set $T > T_c$. Proposition 9.27, (9.111) and (9.106) establish that for $i = 1, 2$,

$$(f(u_3) - f(u_i))(s, \tau) \le p\theta_i^{p-1}(u_3 - u_i)(s, \tau) \tag{9.112}$$

for all $(s, \tau) \in D_{T_c}$, where $\theta_i(s, \tau) \in [u_i(s, \tau), u_3(s, \tau)]$. Combining (9.112) with (9.111) we have, for $i = 1, 2$,

$$(f(u_3) - f(u_i))(s, \tau) \le p((1-p)c\tau)^{(p-1)/(1-p)}(u_3 - u_i)(s, \tau)$$
$$= \frac{p}{c(1-p)\tau}(u_3 - u_i)(s, \tau) \tag{9.113}$$

for all $(s, \tau) \in D_{T_c}$. The Hölder Equivalence Lemma 5.10 gives (for $i = 1, 2$),

$$0 \le (u_3 - u_i)(x, t)$$

$$\le \delta + \frac{1}{\sqrt{\pi}} \int_0^{T_\delta} \int_{-\infty}^{\infty} (f(u_3) - f(u_i)) \left(x + 2\sqrt{t-\tau}\lambda, \tau\right) e^{-\lambda^2} d\lambda d\tau$$

$$+ \frac{1}{\sqrt{\pi}} \int_{T_\delta}^t \int_{-\infty}^{\infty} (f(u_3) - f(u_i)) \left(x + 2\sqrt{t-\tau}\lambda, \tau\right) e^{-\lambda^2} d\lambda d\tau$$

$$\le \delta + \frac{1}{\sqrt{\pi}} \int_0^{T_\delta} \int_{-\infty}^{\infty} (u_3 - u_i)^p \left(x + 2\sqrt{t-\tau}\lambda, \tau\right) e^{-\lambda^2} d\lambda d\tau$$

$$+ \frac{1}{\sqrt{\pi}} \int_{T_\delta}^t \int_{-\infty}^{\infty} \frac{p}{c(1-p)\tau}(u_3 - u_i) \left(x + 2\sqrt{t-\tau}\lambda, \tau\right) e^{-\lambda^2} d\lambda d\tau$$

$$\leq \delta + \frac{1}{\sqrt{\pi}} \int_0^{T_\delta} \int_{-\infty}^{\infty} 2^{p/(1-p)} \delta^p e^{-\lambda^2} d\lambda d\tau$$

$$+ \frac{1}{\sqrt{\pi}} \int_{T_\delta}^t \int_{-\infty}^{\infty} \frac{p}{c(1-p)\tau} \|(u_3 - u_i)(\cdot, \tau)\|_B e^{-\lambda^2} d\lambda d\tau$$

$$\leq \delta \left(1 + \frac{2^{p/(1-p)}}{(1-p)}\right) + \int_{T_\delta}^t \frac{p}{c(1-p)\tau} \|(u_3 - u_i)(\cdot, \tau)\|_B d\tau \qquad (9.114)$$

for all $(x,t) \in \bar{D}_{T_c}^{T_\delta}$, via (9.49), Proposition 9.26, (9.113) and (9.110) respectively. It follows from (9.114) that

$$\|(u_3 - u_i)(\cdot, t)\|_B \leq \delta \left(1 + \frac{2^{p/(1-p)}}{(1-p)}\right) + \int_{T_\delta}^t \frac{p}{c(1-p)\tau} \|(u_3 - u_i)(\cdot, \tau)\|_B d\tau$$

$$(9.115)$$

for all $t \in [T_\delta, T_c]$. Now define $G : [T_\delta, T_c] \to \mathbb{R}^+$ to be

$$G(t) = \delta \left(1 + \frac{2^{p/(1-p)}}{(1-p)}\right) + \int_{T_\delta}^t \frac{p}{c(1-p)\tau} \|(u_3 - u_i)(\cdot, \tau)\|_B d\tau \quad (9.116)$$

for all $t \in [T_\delta, T_c]$. It follows from (9.115), (9.116), Corollary 5.16 and the fundamental theorem of calculus, that G is differentiable on $[T_\delta, T_c]$ and satisfies

$$\frac{1}{G(\tau)} \frac{dG(\tau)}{d\tau} \leq \frac{p}{c(1-p)\tau}; \quad \forall \tau \in [T_\delta, T_c]. \qquad (9.117)$$

Upon integrating both sides of (9.117) with respect to τ from T_δ to $t \in [T_\delta, T_c]$, we obtain

$$\ln \left(\frac{G(t)}{\delta \left(1 + \frac{2^{p/(1-p)}}{(1-p)}\right)}\right) \leq \frac{p}{c(1-p)} \ln \left(\frac{t(1-p)}{\delta^{(1-p)}}\right)$$

$$\leq \ln \left(\frac{(T_c(1-p))^{p/c(1-p)}}{\delta^{p/c}}\right) \qquad (9.118)$$

for all $t \in [T_\delta, T_c]$. Taking exponentials of both sides of (9.118) and re-arranging gives

$$G(t) \leq \delta^{(1-p/c)} \left(1 + \frac{2^{p/(1-p)}}{(1-p)}\right) ((1-p)T_c)^{p/c(1-p)} = k(p,c)\delta^{(1-p/c)}$$

$$(9.119)$$

for all $t \in [T_\delta, T_c]$, with $k(p,c) = \left(1 + \frac{2^{p/(1-p)}}{(1-p)}\right)((1-p)T_c)^{p/c(1-p)}$ which is *independent of* δ. It follows from (9.119), (9.116) and (9.115) that

$$\|(u_3 - u_i)(\cdot, t)\|_B \leq k(p,c)\delta^{(1-p/c)}; \quad \forall t \in [T_\delta, T_c]. \qquad (9.120)$$

It remains to consider $t \in [T_c, T]$. Now, inequality (9.120) gives

$$||(u_3 - u_i)(\cdot, T_c)||_B \le k(p, c)\delta^{(1-p/c)} \quad (i = 1, 2). \tag{9.121}$$

Also, via (9.111), we have

$$u_i(x, T_c) \ge ((1 - p)cT_c)^{1/(1-p)}; \quad \forall x \in \mathbb{R}, \quad (i = 1, 2, 3). \tag{9.122}$$

Since $u_i : \bar{D}_T \to \mathbb{R}$ are solutions to (S-R-D-C-3), then (9.122) and Theorem 3.6 establish that

$$u_i(x, t) \ge ((1 - p)cT_c)^{1/(1-p)} = k'(p, c); \quad \forall (x, t) \in \bar{D}_T^{T_c} \quad (i = 1, 2, 3), \tag{9.123}$$

with $k'(p, c)$ being *independent of* δ. Now consider $\tilde{u}_i : \bar{D}_{T-T_c} \to \mathbb{R}$ ($i = 1, 2, 3$) given by

$$\tilde{u}_i(x, t) = u_i(x, t + T_c); \quad \forall (x, t) \in \bar{D}_{T-T_c}. \tag{9.124}$$

Then, via (9.121),

$$||(\tilde{u}_3 - \tilde{u}_i)(\cdot, 0)||_B \le k(p, c)\delta^{(1-p/c)} \quad (i = 1, 2). \tag{9.125}$$

It now follows from the Hölder Equivalence Lemma 5.10, (9.125), (9.123), (9.106) and use of the mean value theorem (for f on $[0, 1]$ and $(1, \infty)$), with $\eta = \min\{k'(p, c), \frac{1}{2}\gamma\}$, which is *independent of* δ, that

$$0 \le (\tilde{u}_3 - \tilde{u}_i)(x, t) \le k(p, c)\delta^{(1-p/c)}$$
$$+ \frac{1}{\sqrt{\pi}} \int_0^t \int_{-\infty}^{\infty} (f(\tilde{u}_3) - f(\tilde{u}_i)) \left(x + 2\sqrt{t - \tau}\lambda, \tau\right) e^{-\lambda^2} d\lambda d\tau$$
$$\le k(p, c)\delta^{(1-p/c)}$$
$$+ \frac{1}{\sqrt{\pi}} \int_0^t \int_{-\infty}^{\infty} f'(\eta)(\tilde{u}_3 - \tilde{u}_i) \left(x + 2\sqrt{t - \tau}\lambda, \tau\right) e^{-\lambda^2} d\lambda d\tau$$
$$\le k(p, c)\delta^{(1-p/c)} + \int_0^t f'(\eta)||(\tilde{u}_3 - \tilde{u}_i)(\cdot, \tau)||_B d\tau \tag{9.126}$$

for all $(x, t) \in \bar{D}_{T-T_c}$. Hence, via (9.126) and Proposition 5.6, we have ($i = 1, 2$),

$$||(\tilde{u}_3 - \tilde{u}_i)(\cdot, t)||_B \le k(p, c)\delta^{(1-p/c)} + \int_0^t f'(\eta)||(\tilde{u}_3 - \tilde{u}_i)(\cdot, \tau)||_B d\tau$$
$$\le k(p, c)\delta^{(1-p/c)} e^{f'(\eta)(T - T_c)}, \tag{9.127}$$

for all $t \in [0, T - T_c]$. Therefore, via (9.110), (9.120), (9.124) and (9.127), we have $(i = 1, 2)$,

$$||(u_3 - u_i)(\cdot, t)||_B \leq \begin{cases} 2^{1/(1-p)}\delta & ; t \in [0, T_\delta] \\ k(p,c)\delta^{(1-p/c)} & ; t \in [T_\delta, T_c] \\ k(p,c)\delta^{(1-p/c)}e^{f'(\eta)(T-T_c)} & ; t \in [T_c, T], \end{cases} \quad (9.128)$$

where $k(p,c) > 0$, $T_c > 0$ and $\eta > 0$ are all *independent of* δ. Now, given $\epsilon > 0$ we may choose δ sufficiently small in (9.128) to guarantee that $||(u_3 - u_i)(\cdot, t)||_B < \frac{1}{2}\epsilon$ for all $t \in [0, T]$, and hence that $||u_3 - u_i||_A < \frac{1}{2}\epsilon$ for $i = 1, 2$. Thus $||u_2 - u_1||_A < \epsilon$, as required.   □

An immediate consequence of this result is,

Corollary$_\ddagger$ 9.36 *The problem (S-R-D-C-3) is globally well-posed on* $BPC_{+'}^2(\mathbb{R})$.

Proof (P1), (P2) and (P3) follow from Theorem 9.31, Theorem 9.33 and Theorem 9.35 respectively.   □

To establish a uniform global well-posedness result, some additional qualitative information is required.

Proposition$_\ddagger$ 9.37 *For any* $u_0 \in BPC_{+'}^2(\mathbb{R})$, *the corresponding unique solution* $u : \bar{D}_\infty \to \mathbb{R}$ *to (S-R-D-C-3) satisfies*

$$u(x, t) \geq 1; \quad \forall(x, t) \in \bar{D}_\infty^{I_1},$$

where $I_1 = \int_0^1 \frac{1}{r^p(1-r)^q} dr$.

Proof Consider the function $I : [0, 1] \to \mathbb{R}$ given by

$$I(s) = \int_0^s \frac{1}{r^p(1-r)^q} dr; \quad \forall s \in [0, 1], \quad (9.129)$$

where the improper integral is implied (since $p, q \in (0, 1)$). It is readily established that I is continuous and bounded on $[0, 1]$ and differentiable on $(0, 1)$, with derivative given by

$$I'(s) = \frac{1}{s^p(1-s)^q}; \quad \forall s \in (0, 1). \quad (9.130)$$

It follows from (9.130) that I is strictly increasing for all $s \in [0, 1]$, and hence

$$I : [0, 1] \to [0, I_1] \text{ is a bijection.} \quad (9.131)$$

We conclude from (9.130), (9.131) and the inverse function theorem [66] (p.221–222) that there exists a function $J : [0, I_1] \to [0, 1]$ such that

$$J(I(s)) = s \quad \forall s \in [0, 1], \quad I(J(t)) = t; \quad \forall t \in [0, I_1], \quad J(0) = 0, \quad J(I_1) = 1. \tag{9.132}$$

Moreover, J is continuous and increasing on $[0, I_1]$ and differentiable on $[0, I_1]$ with a derivative given by

$$J'(t) = (J(t))^p (1 - J(t))^q; \quad \forall t \in [0, I_1]. \tag{9.133}$$

It follows from (9.133) that J' is continuous and therefore bounded on $[0, I_1]$ with

$$J'(0) = J'(I_1) = 0. \tag{9.134}$$

Now consider $\underline{u} : \bar{D}_\infty \to \mathbb{R}$ given by

$$\underline{u}(x, t) = \begin{cases} J(t) & ; (x, t) \in \bar{D}_{I_1} \\ 1 & ; (x, t) \in D_\infty^{I_1}. \end{cases} \tag{9.135}$$

It follows from (9.132), (9.133), (9.134) and (9.135) that $\underline{u}$ is continuous and bounded on $\bar{D}_\infty$, whilst $\underline{u}_t$, $\underline{u}_x$ and $\underline{u}_{xx}$ exist and are continuous on D_∞. Additionally, for $f : \mathbb{R} \to \mathbb{R}$ given by (9.49), $\underline{u}$ satisfies

$$\underline{u}_t - \underline{u}_{xx} - f(\underline{u}) = 0 \leq 0; \quad \forall (x, t) \in D_\infty, \tag{9.136}$$

$$\underline{u}(x, 0) = 0; \quad \forall x \in \mathbb{R}, \tag{9.137}$$

via (9.133) and (9.132). It follows from (9.135), (9.136) and (9.137) that $\underline{u}$ is a R-S-B to (S-R-D-C-3) on $\bar{D}_T$ (any $T > 0$) for any initial data $u_0 \in \mathrm{BPC}^2_{+'}(\mathbb{R})$. Also with $u : \bar{D}_\infty \to \mathbb{R}$ being the unique solution to (S-R-D-C-3) with corresponding initial data $u_0 \in \mathrm{BPC}^2_{+'}(\mathbb{R})$, we may take u as a R-S-P to (S-R-D-C-3). An application of Corollary 9.34 gives

$$\underline{u}(x, t) \leq u(x, t); \quad \forall (x, t) \in \bar{D}_\infty. \tag{9.138}$$

The result follows from (9.129), (9.135) and (9.138).   □

We can now establish uniform global well-posedness for (S-R-D-C-3). Namely,

Corollary$_\ddagger$ **9.38** *The problem (S-R-D-C-3) is uniformly globally well-posed on* $BPC^2_{+'}(\mathbb{R})$.

Proof (P1) and (P2) follow from Theorem 9.31 and Theorem 9.33 respectively. Also, via Theorem 9.35, for any $u_{10} \in \text{BPC}^2_{+'}(\mathbb{R})$ and any $\epsilon > 0$, there exists $\delta > 0$ such that, for all $u_{20} \in \text{BPC}^2_{+'}(\mathbb{R})$ that satisfy $||(u_{10} - u_{20})||_B < \delta$, then the corresponding solutions $u_1, u_2 : \bar{D}_\infty \to \mathbb{R}$ to (S-R-D-C-3) satisfy

$$||(u_1 - u_2)(\cdot, t)||_B < \epsilon; \quad \forall t \in [0, I_1], \tag{9.139}$$

with I_1 as in Proposition 9.37. Now consider the functions $\tilde{u}_1, \tilde{u}_2 : \bar{D}_\infty \to \mathbb{R}$ given by

$$\tilde{u}_i(x, t) = u_i(x, t + I_1) \quad (i = 1, 2); \quad \forall(x, t) \in \bar{D}_\infty. \tag{9.140}$$

It follows from (9.139), (9.140), Proposition 9.37 and (9.49) that

$$\tilde{u}_{it} - \tilde{u}_{ixx} = 0 \quad (i = 1, 2); \quad \forall(x, t) \in D_\infty, \tag{9.141}$$

$$||(\tilde{u}_1 - \tilde{u}_2)(\cdot, 0)||_B < \epsilon, \tag{9.142}$$

where $(\tilde{u}_1 - \tilde{u}_2)(\cdot, 0) \in \text{BPC}^2_{+'}(\mathbb{R})$ via Proposition 9.37, Proposition 9.25 and Theorem 9.31 with Lemma 5.12 and Lemma 5.15. Therefore, via Theorem 4.2, (9.141) and (9.142), we have

$$||(\tilde{u}_1 - \tilde{u}_2)(\cdot, t)||_B \le \frac{1}{\sqrt{\pi}} \int_{-\infty}^\infty ||(\tilde{u}_1 - \tilde{u}_2)(\cdot, 0)||_B e^{-\lambda^2} d\lambda < \epsilon; \quad \forall t \in [0, \infty). \tag{9.143}$$

It follows from (9.139), (9.140) and (9.143) that for any $u_0 \in \text{BPC}^2_{+'}(\mathbb{R})$, there exists a constant $\delta > 0$, such that for all $u_0' \in \text{BPC}^2_{+'}(\mathbb{R})$ that satisfy $||(u_0 - u_0')||_B < \delta$, then the corresponding solutions $u, u' : \bar{D}_T \to \mathbb{R}$ to (S-R-D-C-3) satisfy $||(u - u')(\cdot, t)||_B < \epsilon$ for all $t \in [0, \infty)$, and hence (P3) is satisfied, as required. $\square$

We conclude by developing some qualitative properties of solutions to (S-R-D-C-3). Firstly, we introduce the functions $w_+, w_- : [0, \infty) \to \mathbb{R}$ such that, with $M_0 \le 1$,

$$w_-(t) = \begin{cases} \phi_-(t) & ; 0 \le t \le t_- \\ 1 & ; t > t_-, \end{cases}$$

$$w_+(t) = \begin{cases} \phi_+(t) & ; 0 \le t \le t_+ \\ 1 & ; t > t_+, \end{cases}$$

where t_+ and t_- are given by

$$t_- = \int_{m_0}^1 \frac{1}{s^p(1-s)^q} ds, \quad t_+ = \int_{M_0}^1 \frac{1}{s^p(1-s)^q} ds, \tag{9.144}$$

and $\phi_+(t)$, $\phi_-(t)$ are defined implicitly by

$$\int_{m_0}^{\phi_-(t)} \frac{1}{s^p(1-s)^q} ds = t, \quad \forall t \in [0, t_-],$$

$$\int_{M_0}^{\phi_+(t)} \frac{1}{s^p(1-s)^q} ds = t, \quad \forall t \in [0, t_+]. \tag{9.145}$$

It follows from (9.144) and (9.145) that $w_+, w_- \in C^1([0, \infty))$, $w_+(t)$ and $w_-(t)$ are non-decreasing with $t \in [0, \infty)$, $w_+(0) = M_0$ and $w_-(0) = m_0$ with $w_+(t) \geq w_-(t)$ for all $t \in [0, \infty)$. We now have,

Theorem‡ 9.39 *Let $u : \bar{D}_\infty \to \mathbb{R}$ be the unique solution to (S-R-D-C-3) with $u_0 \in BPC_{+'}^2(\mathbb{R})$ such that $M_0 \leq 1$, then*

$$w_-(t) \leq u(x, t) \leq w_+(t), \quad \forall(x, t) \in \bar{D}_\infty.$$

Proof This follows immediately from Corollary 9.34, upon taking $\underline{u}, \overline{u} : \bar{D}_\infty \to \mathbb{R}$ such that $\underline{u}(x, t) = w_-(t)$ with $\overline{u}(x, t) = u(x, t)$ and $\overline{u}(x, t) = w_+(t)$ with $\underline{u}(x, t) = u(x, t)$ for all $(x, t) \in \bar{D}_\infty$. $\square$

Corollary‡ 9.40 *Let $u : \bar{D}_\infty \to \mathbb{R}$ be the unique solution to (S-R-D-C-3) with $u_0 \in BPC_{+'}^2(\mathbb{R})$ when $M_0 \leq 1$, then $u(x, t) = 1$ for all $(x, t) \in \bar{D}_\infty^{t-}$.*

Proof Follows from Theorem 9.39. $\square$

We next consider (S-R-D-C-3) when $u_0 \in BPC_{+'}^2(\mathbb{R})$ is such that $M_0 > 1$ and $m_0 \leq 1$, with $\mathcal{S}_+ = \{x \in \mathbb{R} : u_0(x) > 1\}$ being bounded. We introduce $U^+ : \bar{D}_\infty \to \mathbb{R}$, such that

$$U^+(x, t) = \frac{1}{\sqrt{\pi}} \int_{-\infty}^{\infty} u_0^+ (x + 2\sqrt{t}\lambda) e^{-\lambda^2} d\lambda; \quad \forall(x, t) \in \bar{D}_\infty, \tag{9.146}$$

with $u_0^+ : \mathbb{R} \to \mathbb{R}$ given by

$$u_0^+(x) = \begin{cases} u_0(x) & ; x \in \mathcal{S}_+ \\ 1 & ; x \in \mathbb{R}\backslash\mathcal{S}_+. \end{cases} \tag{9.147}$$

It follows from (9.146) and (9.147) that.

- U^+ is continuous on $\bar{D}_\infty$, and U_t^+, U_x^+ and U_{xx}^+ exist and are continuous on D_∞.
- $U_t^+ = U_{xx}^+$ on D_∞.

- $U^+(x,t) \to 1$ as $|x| \to \infty$ uniformly for $t \in [0, \infty)$.
- $1 < U^+(x,t) < 1 + \frac{L(M_0-1)}{\sqrt{\pi t}}$ for all $(x,t) \in D_\infty$ where $L = \sup_{\lambda \in S_+} |\lambda|$.

We now have,

Theorem‡ 9.41 *Let $u : \bar{D}_\infty \to \mathbb{R}$ be the unique solution to (S-R-D-C-3) with $u_0 \in BPC^2_{+'}(\mathbb{R})$ when $M_0 > 1$, $m_0 < 1$ and $\mathcal{S}_+$ is bounded. Then,*

$$w_-(t) \le u(x,t) \le U^+(x,t); \quad \forall (x,t) \in \bar{D}_\infty,$$

and

$$1 \le u(x,t) < 1 + \frac{L(M_0-1)}{\sqrt{\pi t}}; \quad \forall (x,t) \in \bar{D}^t_\infty.$$

Proof Follows from Corollary 9.34 and the properties of U^+ and w_- established above. $\qquad\square$

10

Extensions and Concluding Remarks

The remarks which conclude this monograph are split into three sections. In the first section we comment on the problem (B-R-D-C) when allowing for more general classes of initial data and illustrate the associated modifications required for the theory to be developed in the same way as in the preceding chapters. The second section covers ideas that are extensions of the work contained within the monograph and which appear obtainable, via methods within the monograph. In the third section, two ideas are introduced, for which a method of proof is most likely not contained in this monograph. Nonetheless, these ideas have arisen as a result of this monograph and any development in this area would be of significant interest.

10.1 Extensions

To begin this section we consider two extended initial data sets for which we anticipate that the majority of the theory contained in this monograph is directly applicable. Consider $u_0 \in B_B$ and define B_{B+} to be

$$B_{B+} = \{u \in B_B : u(x) \geq 0; \ \forall x \in \mathbb{R}\},$$

with B_B and B_{B+} now playing an analogous role to $\mathrm{BPC}^2(\mathbb{R})$ and $\mathrm{BPC}^2_+(\mathbb{R})$, respectively, in what follows. We now illustrate the required modifications so that the theory developed in Chapters 2–8 will apply to the problem (B-R-D-C) with initial data $u_0 \in B_B$. Now, replacing $\mathrm{BPC}^2(\mathbb{R})$ with B_B in Chapter 2 and Chapter 3 requires no modification at all. However, in Chapter 4, since u_0 is now not necessarily differentiable, specific results must be modified; namely, equations (4.10), (4.11) and (4.12). This arises due to an inability to integrate by parts (since u_0 is not necessarily differentiable), and hence, upon differentiating (4.8), and making an appropriate substitution, the derivatives of (4.8) are now given by the following expressions,

$$u_x(x,t) = \frac{1}{2\sqrt{\pi t}} \int_{-\infty}^{\infty} u_0(s) \left(-\frac{(x-s)}{2t} \right) e^{-\frac{(x-s)^2}{4t}} ds$$

$$= \frac{1}{\sqrt{\pi t}} \int_{-\infty}^{\infty} u_0 \left(x + 2\sqrt{t}w \right) w e^{-w^2} dw \qquad (10.1)$$

$$u_{xx}(x,t) = \frac{-1}{4\sqrt{\pi}t^{3/2}} \int_{-\infty}^{\infty} u_0(s) e^{-\frac{(x-s)^2}{4t}} ds$$

$$+ \frac{1}{2\sqrt{\pi}t^{3/2}} \int_{-\infty}^{\infty} u_0(s) \left(\frac{(x-s)^2}{4t} \right) e^{-\frac{(x-s)^2}{4t}} ds$$

$$= \frac{1}{\sqrt{\pi t}} \int_{-\infty}^{\infty} u_0 \left(x + 2\sqrt{t}w \right) (w^2 - 1/2) e^{-w^2} dw \qquad (10.2)$$

$$u_t(x,t) = \frac{-1}{4\sqrt{\pi}t^{3/2}} \int_{-\infty}^{\infty} u_0(s) e^{-\frac{(x-s)^2}{4t}} ds$$

$$+ \frac{1}{2\sqrt{\pi}t^{3/2}} \int_{-\infty}^{\infty} u_0(s) \left(\frac{(x-s)^2}{4t} \right) e^{-\frac{(x-s)^2}{4t}} ds$$

$$= \frac{1}{\sqrt{\pi t}} \int_{-\infty}^{\infty} u_0 \left(x + 2\sqrt{t}w \right) (w^2 - 1/2) e^{-w^2} dw \qquad (10.3)$$

for all $(x,t) \in D_T$. It thus follows from (10.1), (10.2) and (10.3) that $u : \bar{D}_T \to \mathbb{R}$ given by (4.8) satisfies

$$|u_x(x,t)| \le \frac{||u_0||_B}{\sqrt{\pi t}}, \quad |u_{xx}(x,t)| \le \frac{||u_0||_B}{t}, \quad |u_t(x,t)| \le \frac{||u_0||_B}{t}, \qquad (10.4)$$

for all $(x,t) \in D_T$. This leads to an alternative version of Theorem 4.7, namely,

Theorem 10.1 (Bounds and Derivative Estimates) *Let $u : \bar{D}_\infty \to \mathbb{R}$ be the unique solution to (B-D-C) on $\bar{D}_\infty$ given by (4.8) for $u_0 \in B_B$. Then,*

$$|u(x,t)| \le ||u_0||_B, \quad |u_x(x,t)| \le \frac{||u_0||_B}{\sqrt{\pi t}},$$

$$|u_{xx}(x,t)| \le \frac{||u_0||_B}{t}, \quad |u_t(x,t)| \le \frac{||u_0||_B}{t}, \qquad (10.5)$$

for all $(x,t) \in D_\infty$.

Now, in Chapter 5, to obtain Lemma 5.10, we must re-work Section 5.1 and Section 5.2 with the modified condition (H') instead of (H) on $F : \bar{D}_T \to \mathbb{R}$, namely,

(H') $F : \bar{D}_T \to \mathbb{R}$ is continuous, bounded and uniformly Hölder continuous of degree $0 < \alpha \le 1$ with respect to $x \in \mathbb{R}$ for $t \in (0, T]$. Moreover, there exist positive constants k_T^1 and k_T^2, and $\beta \in (0, 1)$ (independent of $t \in (0, T]$) such that

$$|F(y,t) - F(x,t)| \leq \left(\frac{k_T^1}{t^\beta} + k_T^2\right)|y - x|^\alpha, \quad \forall(y,t),(x,t) \in D_T.$$

Then Theorem 5.1 remains unaltered, whilst Theorem 5.2 acquires derivative bounds, which can now (depending on β) blow up as $t \to 0^+$. The results in Section 5.2 follow without modification. The proofs of these modified results largely follow the same steps as the originals; however, there is an additional step which is required, which is illustrated in the forthcoming proof of the modified derivative estimate lemma of Section 5.4. In Chapter 5, Section 3, it follows, with the above modifications, that both Lemma 5.10 and Lemma 5.11 continue to hold for $u_0 \in B_B$ replacing $u_0 \in \text{BPC}^2(\mathbb{R})$. Now, since Chapter 5, Section 4 is comprised of derivative estimates (which imply bounds on the derivatives of solutions to (B-R-D-C) on $\bar{D}_T$), changes are required. Given below are statements of the modified versions of Lemma 5.12, Lemma 5.14 and Lemma 5.15 respectively. The proofs of these modified results follow the same steps as in the corresponding proofs in Chapter 5, except that of Lemma 5.15 which requires an additional calculation.

Lemma 10.2 (Derivative Estimate) *Let* $f \in H_\alpha$ *for some* $\alpha \in (0,1]$ *and* $u : \bar{D}_T \to \mathbb{R}$ *be a solution to (B-R-D-C) with* $u_0 \in B_B$, *on* $\bar{D}_T$. *Then,*

$$|u_x(x,t)| \leq \frac{2M_T}{\sqrt{\pi}}(1 + T^{\frac{1}{2}}) + \frac{||u_0||_B}{\sqrt{\pi t}}, \quad \forall(x,t) \in D_T,$$

where $M_T > 0$ *is an upper bound for* $|f \circ u| : \bar{D}_T \to \mathbb{R}$.

Lemma 10.3 *Let* $f \in H_\alpha$ *for some* $\alpha \in (0,1]$ *and let* $u : \bar{D}_T \to \mathbb{R}$ *be a solution to (B-R-D-C) with* $u_0 \in B_B$, *on* $\bar{D}_T$. *Then* $f \circ u : \bar{D}_T \to \mathbb{R}$ *satisfies*

$$|f(u(y,t)) - f(u(x,t))| \leq k_T(t)|y - x|^\alpha; \quad \forall(x,t),(y,t) \in D_T,$$

where $k_T : (0,T] \to \mathbb{R}$ *is given by*

$$k_T(t) = k_E\left(\frac{2M_T}{\sqrt{\pi}}(1 + T^{\frac{1}{2}}) + \frac{||u_0||_B}{\sqrt{\pi t}}\right)^\alpha; \quad \forall t \in (0,T]$$

and $k_E > 0$ *is a Hölder constant for* $f : \mathbb{R} \to \mathbb{R}$ *on the closed bounded interval* $[-U_T, U_T]$, *with* $U_T > 0$ *being an upper bound for* $|u| : \bar{D}_T \to \mathbb{R}$.

Lemma 10.4 (Derivative Estimates) *Let* $f \in H_\alpha$ *for some* $\alpha \in (0,1]$ *and* $u : \bar{D}_T \to \mathbb{R}$ *be a solution to (B-R-D-C) with* $u_0 \in B_B$, *on* $\bar{D}_T$. *Then,*

$$|u_{xx}(x,t)| \le K(\alpha, T, u_0, M_T, k_E) + \frac{\|u_0\|_B}{t}; \quad \forall (x,t) \in D_T,$$

$$|u_t(x,t)| \le K(\alpha, T, u_0, M_T, k_E) + \frac{\|u_0\|_B}{t} + M_T; \quad \forall (x,t) \in D_T,$$

where

$$K(\alpha, T, u_0, M_T, k_E) = \frac{2^{\alpha+1} I_\alpha k_E}{\alpha \sqrt{\pi}} \left(\frac{2M_T}{\sqrt{\pi}} (1 + T^{\frac{1}{2}}) \right)^\alpha (1+T)^{\alpha/2}$$
$$+ \frac{2^{\alpha+2} I_\alpha k_E \|u_0\|_B^\alpha}{\pi^{(\alpha+1)/2} \alpha (2-\alpha)},$$

and

$$I_\alpha = \int_{-\infty}^{\infty} |\lambda|^\alpha |\lambda^2 - 1/2| e^{-\lambda^2} d\lambda \quad (> 0).$$

The beginning of the proof of Lemma 10.4 follows the same steps as that of Lemma 5.15 until line (5.70), which now becomes, for $n \in \mathbb{N}$ such that $1/n < \delta/2 \le t/2$ (for all $t \in [\delta, T]$), we have

$$\frac{1}{\sqrt{\pi}} \left| \int_0^{t-1/n} \int_{-\infty}^{\infty} \frac{f(u(x + 2\sqrt{t-\tau}\lambda, \tau))}{(t-\tau)} (\lambda^2 - 1/2) e^{-\lambda^2} d\lambda d\tau \right|$$

$$\le \frac{2^\alpha}{\sqrt{\pi}} \int_{1/n}^{t-1/n} \int_{-\infty}^{\infty} \frac{|\lambda|^\alpha |\lambda^2 - 1/2| k_T(\tau)}{(t-\tau)^{1-\alpha/2}} e^{-\lambda^2} d\lambda d\tau \quad \text{(via Lemma 10.3)}$$

$$+ \frac{1}{\sqrt{\pi}} \int_0^{1/n} \int_{-\infty}^{\infty} \frac{M_T}{(t - 1/n)} |\lambda^2 - 1/2| e^{-\lambda^2} d\lambda d\tau$$

$$\le \frac{2^\alpha I_\alpha k_E}{\sqrt{\pi}} \int_{1/n}^{t-1/n} \left(\left(\frac{2M_T}{\sqrt{\pi}}(1 + T^{\frac{1}{2}}) \right)^\alpha \frac{1}{(t-\tau)^{1-\alpha/2}} \right.$$
$$\left. + \left(\frac{\|u_0\|_B}{\sqrt{\pi\tau}} \right)^\alpha \frac{1}{(t-\tau)^{1-\alpha/2}} \right) d\tau + \frac{M_T \hat{I}}{n\sqrt{\pi}(t - 1/n)}$$

$$\le \frac{2^{\alpha+1} I_\alpha k_E}{\alpha \sqrt{\pi}} \left(\frac{2M_T}{\sqrt{\pi}}(1 + T^{\frac{1}{2}}) \right)^\alpha (1+T)^{\alpha/2}$$

$$+ \frac{2^\alpha I_\alpha k_E \|u_0\|_B^\alpha}{\pi^{(\alpha+1)/2}} \left(\frac{2}{t} \right)^{1-\alpha/2} \int_{1/n}^{t/2} \frac{1}{\tau^{\alpha/2}} d\tau$$

$$+ \frac{2^\alpha I_\alpha k_E \|u_0\|_B^\alpha}{\pi^{(\alpha+1)/2}} \left(\frac{2}{t} \right)^{\alpha/2} \int_{t/2}^{t-1/n} \frac{1}{(t-\tau)^{1-\alpha/2}} d\tau + \frac{M_T \hat{I}}{n\sqrt{\pi}(t - 1/n)}$$

$$\le \frac{2^{\alpha+1} I_\alpha k_E}{\alpha \sqrt{\pi}} \left(\frac{2M_T}{\sqrt{\pi}}(1 + T^{\frac{1}{2}}) \right)^\alpha (1+T)^{\alpha/2}$$

$$+ \frac{2^{\alpha+1} I_\alpha k_E \|u_0\|_B^\alpha}{\pi^{(\alpha+1)/2}(2-\alpha)} + \frac{2^{\alpha+1} I_\alpha k_E \|u_0\|_B^\alpha}{\pi^{(\alpha+1)/2}\alpha} + \frac{M_T \hat{I}}{n\sqrt{\pi}(t-1/n)}$$

$$= K(\alpha, T, u_0, M_T, k_E) + \frac{M_T \hat{I}}{n\sqrt{\pi}(t-1/n)}, \tag{10.6}$$

where

$$\hat{I} = \int_{-\infty}^{\infty} |\lambda^2 - 1/2| e^{-\lambda^2} d\lambda.$$

It thus follows from the corresponding line to (5.68) with (10.6) that

$$|u_{xx}(x,t)| \le \frac{\|u_0\|_B}{t} + K(\alpha, T, u_0, M_T, k_E); \quad \forall (x,t) \in D_T,$$

as required. The result for u_t follows similarly.

We conclude from the above discussion that when $u_0 \in B_B$, then a weaker version of Corollary 5.16 is available, namely,

Corollary 10.5 *Let $f \in H_\alpha$ for some $\alpha \in (0,1]$ and $u : \bar{D}_T \to \mathbb{R}$ be a solution to (B-R-D-C) with $u_0 \in B_B$, on $\bar{D}_T$. Then u is uniformly continuous on $\bar{D}_T^\delta$ for any $\delta \in (0,T)$.*

Next, we consider modifications of the results in Chapters 6–8. To begin, we observe that now there is no guarantee of uniform continuity of solutions to (B-R-D-C) on D_T when $u_0 \in B_B$. Thus the step in the proof of Theorem 6.7 which guarantees that

$$\|u_1(\cdot, t) - u_2(\cdot, t)\|_B$$

is continuous no longer follows. However, we can apply Lemma 5.5 to establish that

$$\|u_1(\cdot, t) - u_2(\cdot, t)\|_B \in L^1([0, T]),$$

which is all that is required to then apply Proposition 5.6. The rest of Chapter 6 needs no additional modifications, whilst Chapter 7 needs no modification at all. However, in Chapter 8, the following modifications must be made to accommodate initial data $u_0 \in B_B$. The first of these changes stems from Lemma 10.2, Lemma 10.3 and Lemma 10.4, and replaces Proposition 8.14, namely,

Proposition 10.6 *Let $\underline{u}_n : \bar{D}_\delta \to \mathbb{R}$ be the (unique) solution to (B-R-D-C)$_n^l$ ($n \in \mathbb{N}$). Then, on D_δ, we have*

$$|\underline{u}_{nx}(x,t)| \le \frac{2c'}{\sqrt{\pi}}(1 + \delta^{1/2}) + \frac{\|u_0\|_B}{\sqrt{\pi t}},$$

$$|\underline{u}_{nxx}(x,t)| \leq \frac{||u_0||_B}{t} + K(\alpha, \delta, u_0, c', 3k_H),$$

$$|\underline{u}_{nt}(x,t)| \leq \frac{||u_0||_B}{t} + K(\alpha, \delta, u_0, c', 3k_H) + c'$$

for all $(x,t) \in D_\delta$, *where* $k_H > 0$ *is a Hölder constant for* $f \in H_\alpha$ *on*

$$[-(||u_0||_B + 1), (||u_0||_B + 1)].$$

The proof of this result follows directly from Lemma 10.2, Lemma 10.3 and Lemma 10.4. Next, the proof of Lemma 8.19, when $u_0 \in B_B$, is slightly more complicated. To prove the result when $u_0 \in B_B$, we first replace $\bar{D}_\delta^{0,X}$ with $\bar{D}_\delta^{Y,X}$ for any $X > 0$ and $\delta > Y > 0$. We now follow the steps of the proof and using the Proposition 10.6 we conclude that u^* is continuous on D_δ and is bounded on $\bar{D}_\delta$. However, to establish that u^* is continuous on $\bar{D}_\delta$, we require u^* to be continuous when $(x,t) \in \partial D$. To this end, we first observe that since $\underline{u}_n(x,0) = u_0(x) - \frac{1}{2n}$ for all $x \in \mathbb{R}$ and $n \in \mathbb{N}$, then

$$u^*(x,0) = u_0(x), \quad \forall x \in \mathbb{R}.$$

Next take $\epsilon > 0$. Now, for any $h_1 \in \mathbb{R}$ and $h_2 \in (0, \delta]$, we have

$$|u^*(x,0) - u^*(x+h_1, h_2)| = |u_0(x) - \lim_{j \to \infty} u_{nj}(x+h_1, h_2)|, \quad (10.7)$$

for some sequence $\{u_{nj}(x+h_1, h_2)\}_{n_j \in \mathbb{N}}$ given in the construction. Now take

$$h_2 < \min\left\{\frac{\epsilon}{4c'}, \delta\right\}$$

and it follows from Lemma 5.10, (10.7) and the definition of $c' > 0$, that

$$|u^*(x,0) - u^*(x+h_1, h_2)|$$
$$\leq \left| u_0(x) - \frac{1}{\sqrt{\pi}} \int_{-\infty}^{\infty} u_0\left(x + h_1 + 2\sqrt{h_2}\lambda\right) e^{-\lambda^2} d\lambda \right|$$
$$+ \lim_{j \to \infty} \left| \frac{1}{\sqrt{\pi}} \int_0^{h_2} \int_{-\infty}^{\infty} \underline{f}_{n_j}\left(\underline{u}_{n_j}\left(x + h_1 + 2\sqrt{h_2 - \tau}\lambda, \tau\right)\right) e^{-\lambda^2} d\lambda d\tau \right|$$
$$\leq \frac{1}{\sqrt{\pi}} \int_{-\infty}^{\infty} |u_0(x) - u_0\left(x + h_1 + 2\sqrt{h_2}\lambda\right)| e^{-\lambda^2} d\lambda + \frac{\epsilon}{4} \quad (10.8)$$

for any $h_1 \in \mathbb{R}$ and $h_2 < \min\left\{\frac{\epsilon}{4c'}, \delta\right\}$. Next, set

$$\lambda_\epsilon = \max\left\{1, \frac{8||u_0||_B}{\epsilon}\right\}.$$

It then follows from (10.8) that

$$|u^*(x, 0) - u^*(x + h_1, h_2)|$$

$$\leq \int_{-\infty}^{-\lambda_\epsilon} \frac{2||u_0||_B}{\lambda^2} d\lambda + \int_{\lambda_\epsilon}^{\infty} \frac{2||u_0||_B}{\lambda^2} d\lambda$$

$$+ \int_{-\lambda_\epsilon}^{\lambda_\epsilon} |u_0(x) - u_0(x + h_1 + 2\sqrt{h_2}\lambda)| e^{-\lambda^2} d\lambda + \frac{\epsilon}{4}$$

$$\leq \frac{3\epsilon}{4} + \int_{-\lambda_\epsilon}^{\lambda_\epsilon} |u_0(x) - u_0(x + h_1 + 2\sqrt{h_2}\lambda)| e^{-\lambda^2} d\lambda \qquad (10.9)$$

for any $h_1 \in \mathbb{R}$ and $h_2 < \min\{\frac{\epsilon}{4c'}, \delta\}$. Now, since $u_0 \in B_B$, then there exists $\delta_1 > 0$ (which may depend on x) such that

$$|u_0(x) - u_0(y)| < \frac{\epsilon}{8\lambda_\epsilon} \qquad (10.10)$$

for all $y \in \mathbb{R}$ such that $|x - y| < \delta_1$. Thus, for

$$0 < |h_1| < \frac{\delta_1}{2}, \quad 0 < h_2 < \min\left\{\left(\frac{\delta_1}{4\lambda_\epsilon}\right)^2, \frac{\epsilon}{4c'}, \delta\right\},$$

it follows from (10.9) and (10.10) that

$$\int_{-\lambda_\epsilon}^{\lambda_\epsilon} |u_0(x) - u_0(x + h_1 + 2\sqrt{h_2}\lambda)| e^{-\lambda^2} d\lambda \leq \int_{-\lambda_\epsilon}^{\lambda_\epsilon} \frac{\epsilon}{8\lambda_\epsilon} d\lambda = \frac{\epsilon}{4}. \qquad (10.11)$$

It therefore follows from (10.11) and (10.9) that for $h_1, h_2 \in \mathbb{R}$ that satisfy

$$0 < |h_1| < \frac{\delta_1}{2}, \quad 0 < h_2 < \min\left\{\left(\frac{\delta_1}{4\lambda_\epsilon}\right)^2, \frac{\epsilon}{4c'}, \delta\right\},$$

then

$$|u^*(x, 0) - u^*(x + h_1, h_2)| < \epsilon.$$

It thus follows that u^* is continuous on $\bar{D}_\delta$, which completes the proof of the result corresponding to Lemma 8.19. The remaining result which requires attention is Lemma 8.23. The proof follows in a similar fashion, except an additional truncation must be made in the integral in (8.45); specifically

$$\frac{1}{\sqrt{\pi}} \int_0^t \int_{-\lambda_\epsilon}^{\lambda_\epsilon} |\underline{f}_n\left(\underline{u}_n\left(x + 2\sqrt{t - \tau}\lambda, \tau\right)\right)$$

$$- f\left(u^*\left(x + 2\sqrt{t - \tau}\lambda, \tau\right)\right)| e^{-\lambda^2} d\lambda d\tau$$

$$= \frac{1}{\sqrt{\pi}} \int_0^{t'} \int_{-\lambda_\epsilon}^{\lambda_\epsilon} |\underline{f}_n\left(\underline{u}_n\left(x + 2\sqrt{t - \tau}\lambda, \tau\right)\right)$$

$$- f\left(u^*\left(x + 2\sqrt{t - \tau}\lambda, \tau\right)\right)|e^{-\lambda^2}d\lambda d\tau$$

$$+ \frac{1}{\sqrt{\pi}}\int_{t'}^{t}\int_{-\lambda_\epsilon}^{\lambda_\epsilon}|\underline{f}_n\left(\underline{u}_n\left(x + 2\sqrt{t - \tau}\lambda, \tau\right)\right)$$

$$- f\left(u^*\left(x + 2\sqrt{t - \tau}\lambda, \tau\right)\right)|e^{-\lambda^2}d\lambda d\tau$$

$$\leq \frac{\epsilon}{4} + \frac{1}{\sqrt{\pi}}\int_{t'}^{t}\int_{-\lambda_\epsilon}^{\lambda_\epsilon}|\underline{f}_n\left(\underline{u}_n\left(x + 2\sqrt{t - \tau}\lambda, \tau\right)\right)$$

$$- f\left(u^*\left(x + 2\sqrt{t - \tau}\lambda, \tau\right)\right)|e^{-\lambda^2}d\lambda d\tau$$

for sufficiently small $t' > 0$. Now, upon using the corresponding version of Lemma 8.19 with $u_0 \in B_B$, as in the original proof, we can force the remaining term to be less than $\epsilon/4$ (instead of $\epsilon/2$) to complete the proof of the corresponding version of Lemma 8.23. Nothing else in Chapter 8 requires any additional comment.

The only modifications required in Chapter 9 concern the use of Corollary 5.16. These are either dealt with as in Theorem 6.7 in an application of Lemma 5.5, or, in certain cases, the sets concerned need to be modified.

10.2 Possible Extensions

In this section, we consider extensions for which the methodology developed in this monograph may be of use, but will require additional theory from other sources. To begin, consider the problem (B-R-D-C) as stated in Chapter 2, but with the domain $D_T = \mathbb{R} \times (0, T]$ replaced by $D_T = \mathbb{R}^n \times (0, T]$ where we now write $x = (x_1, x_2, \ldots, x_n)$ and equations (2.1) and (2.2) are replaced by

$$u_t - \Delta u = f(u); \quad \forall(x, t) \in D_T, \tag{10.12}$$

$$u(x, 0) = u_0(x); \quad \forall x \in \mathbb{R}^n. \tag{10.13}$$

It is expected that much of the theory contained in this monograph is applicable to this more general problem. In particular, the maximum principles in Chapter 3 extend, without any additional technicalities, when we replace the differential inequalities (3.36) and (3.44) with a differential inequality of the form

$$u_t - \Delta u - h(x, t)u \leq 0; \quad \forall(x, t) \in D_T.$$

The results in Chapter 4 are readily extended upon considering the unique solution to the diffusion equation on $\mathbb{R}^n$ with "smooth enough" initial data $u_0 : \mathbb{R}^n \to \mathbb{R}$ (following Section 10.1, at least continuous and bounded initial data), given by

$$u(x,t) = \begin{cases} \dfrac{1}{(4\pi t)^{n/2}} \displaystyle\int_{\mathbb{R}^n} u_0(s) e^{-\frac{|x-s|^2}{4t}}\, ds & ; (x,t) \in \mathbb{R}^n \times (0,T] \\[2ex] u_0(x) & ; (x,t) \in \mathbb{R}^n \times \{0\}, \end{cases}$$

where $\int_{\mathbb{R}^n} = \int_{s_1=-\infty}^{\infty} \cdots \int_{s_n=-\infty}^{\infty}$ and $ds = ds_n, \cdots, ds_1$. Chapter 5 can be developed in a similar way upon considering

$$\phi(x,t) = \frac{1}{\pi^{n/2}} \int_0^t \int_{\mathbb{R}^n} F\left(x + 2\sqrt{t-\tau}\, s, \tau\right) e^{-s^2}\, ds d\tau, \ \forall (x,t) \in \mathbb{R}^n \times [0,T],$$

as a replacement for (5.2) with $F : \mathbb{R}^n \times [0,T] \to \mathbb{R}$. Along with having to consider derivatives u_{x_i} and $u_{x_i x_j}$ for $i, j = 1, \ldots, n$, we expect that this chapter can be developed in the same manner for the problem above. Since Chapters 6–8 are primarily concerned with the nonlinear terms, the methodology behind the results should not require any significant additional modifications to be applied to the problem above.

Secondly, consider the problem of finding a classical solution to the initial value problem on $\mathbb{R} \times [0,T]$ for some $T > 0$, with continuous and bounded initial data $u_0 \in \mathbb{R}$ and satisfying the partial differential equation

$$u_t - a(x,t)u_{xx} - b(x,t)u_x = f(x,t,u),$$

on $\mathbb{R} \times (0,T]$, where $a, b : \mathbb{R} \times (0,T] \to \mathbb{R}$ are bounded, continuous and locally Hölder continuous in x, uniformly with respect to $t \in (0,T]$, $a(x,t) \geq A > 0$ for all $(x,t) \in \mathbb{R} \times (0,T]$, and $f : \mathbb{R} \times (0,T] \times \mathbb{R} \to \mathbb{R}$ is bounded, continuous and locally Hölder continuous in x uniformly with respect to $t \in (0,T]$, and locally Hölder continuous in u uniformly with respect to $(x,t) \in \mathbb{R} \times (0,T]$. Now, via methods contained in [21], we expect that a local existence result can be stated for this problem, which requires the above conditions and the additional condition, specifically, that f is locally Lipschitz continuous in u uniformly with respect to $(x,t) \in \mathbb{R} \times (0,T]$. We anticipate that by combining the methods developed in this monograph with the methods contained in [21], a local existence result can be obtained with simply a Hölder condition on all the functions a, b and f. The concept of maximal and minimal solutions is expected to be relevant to this problem.

The third extension is closely related to this monograph and a specific case is given by (1.1)–(1.3). Specifically, consider the initial value problem with initial data $u_0 : \mathbb{R} \to \mathbb{R}^m$, for $m \in \mathbb{N}$, which satisfies the system of reaction-diffusion equations

$$\frac{\partial u_i}{\partial t} - D_i \frac{\partial^2 u_i}{\partial x^2} = f_i(u)$$

on $\mathbb{R} \times (0, T]$ where $D_i > 0$ for each $i = 1, \ldots, m$, and each $f_i : \mathbb{R}^m \to \mathbb{R}$ satisfies a local Hölder condition corresponding to that in Definition 2.5 for $f : \mathbb{R} \to \mathbb{R}$. A solution to this problem is defined as in Definition 2.1. The principle difference between this problem and (B-R-D-C) is that comparison theorems are not as easy to obtain. Only for specific types of problems, where the nonlinear term satisfies specific structures, amounting to significantly more than being locally Lipschitz continuous, can comparison results be obtained. A specific example where it is possible to establish a comparison theorem relating to the problem above is when each f_i is non-decreasing in each variable and locally Lipschitz continuous. It is expected that much of the theory in this monograph extends to this problem, specifically the existence results, albeit the concept of maximal and minimal solutions needs to be broadened. Since we are yet to encounter anything in the literature regarding this subject (it would not be a surprise to discover that there is an analogous concept of such solutions within the framework of dynamical systems), we have introduced the terminology "extremal" or "bounding" solutions. In the case where $m = 1$, these are precisely the maximal and minimal solutions constructed in this monograph. However, when $m > 1$, there are additional types of "extremal" or "bounding" solutions. For example, let $m = 2$. Suppose that $u : \mathbb{R} \times [0, T] \to \mathbb{R}^2$ is a solution to the initial value problem. Moreover, suppose that $\hat{u} : \mathbb{R} \times [0, T] \to \mathbb{R}^2$ is any other solution to the same initial value problem. Then if it is necessary that

$$u_1(x, t) \geq \hat{u}_1(x, t), \quad u_2(x, t) \leq \hat{u}_2(x, t); \quad \forall (x, t) \in \mathbb{R} \times [0, T],$$

then the first and second components of u satisfy a similar condition to that of a maximal solution and a minimal solution for (B-R-D-C) respectively. This motivates the following question, namely, when can we guarantee that there exists an "extremal" or "bounding" solution to the initial value problem above, and moreover, of which type? The use of these solutions is made evident by Theorem 9.33, when considering uniqueness arguments.

10.3 Additional Questions

In this section, several outstanding questions, which have arisen as a result of the study in this monograph, are stated. These questions are related to the monograph, but it appears that an approach to answering them will most likely require additional ideas not contained within this monograph.

In Chapter 8, we obtained a local existence result, namely Theorem 8.3, for the problem (B-R-D-C). However, for continuous $f : \mathbb{R} \to \mathbb{R}$ such that

$f \notin H_\alpha$ for any $\alpha \in (0, 1]$, it is not clear whether there exists a local classical solution to (B-R-D-C), or not. Thus, it makes sense to ask the following question,

Question 10.7 *Consider (B-R-D-C) with reaction function $f : \mathbb{R} \to \mathbb{R}$ which satisfies $f(u) = G(u)$ for all $u \in \mathbb{R}$, where $G : \mathbb{R} \to \mathbb{R}$ is given by (2.10) and initial data $u_0 \in BPC^2(\mathbb{R})$ such that*

$$\inf_{x \in \mathbb{R}} \{u_0(x)\} = 0, \quad \sup_{x \in \mathbb{R}} \{u_0(x)\} = 1.$$

Does this (B-R-D-C) have a classical solution on $\bar{D}_T$ for some $T > 0$?

It follows that since $f \in L_u$, the theory developed in Chapter 7 can be applied to this problem, albeit only when solutions have been found. We note that the existence theory developed in this monograph relies critically on the equivalence lemma (Lemma 5.10), which itself, relies on the Schauder estimates (Lemma 5.15). If the Schauder estimates could be extended to include some $f \notin H_\alpha$ for any $\alpha \in (0, 1]$, then an equivalence lemma, and additionally, an existence result could potentially be obtained. Unfortunately, little more can be said about this problem currently.

Regarding uniqueness, the following observation has been made. For (B-R-D-C) with $f \in H_\alpha$ and $u_0 \in BPC^2(\mathbb{R})$ for which $u_0(x) = 0$ for all $x \in \mathbb{R}$, then there are $f \in H_\alpha$ for which there exist distinct solutions to (B-R-D-C) (see Example 8.28). However, whether or not there exist examples of (B-R-D-C) problems for which non-unique solutions $u_1 : \bar{D}_T \to \mathbb{R}$ and $u_2 : \bar{D}_T \to \mathbb{R}$ exist, where at no time $t \in [0, T]$ is either solution a constant function, is not clear. To this end, we have the following question,

Question 10.8 *Does there exist a (B-R-D-C) problem with $f \in H_\alpha$ for some $\alpha \in (0, 1)$, and initial data $u_0 \in BPC^2(\mathbb{R})$ such that*

$$\inf_{x \in \mathbb{R}} \{u_0(x)\} \neq \sup_{x \in \mathbb{R}} \{u_0(x)\},$$

for which there exist distinct classical solutions $u_1 : \bar{D}_T \to \mathbb{R}$ and $u_2 : \bar{D}_T \to \mathbb{R}$ for some $T > 0$, which satisfy, for $i = 1, 2$,

$$\inf_{x \in \mathbb{R}} \{u_i(x, t)\} \neq \sup_{x \in \mathbb{R}} \{u_i(x, t)\}; \quad \forall t \in [0, T]?$$

Following up on the observation above, when (B-R-D-C) problems with distinct solutions have been identified, they share a particular quality. Namely, that if $f \in H_\alpha$ is not Lipschitz continuous in some closed neighborhood of a point where $f(u) = 0$, then this can give rise to distinct solutions (see Example 8.28 and consider the point $u = 0$). Regarding this point, the following question is significant,

Question 10.9 *Consider (B-R-D-C) with $f \in H_\alpha$, which satisfies*

$$f(u) \geq m_f > 0, \quad \forall u \in \mathbb{R}$$

for some constant $m_f > 0$, with initial data $u_0 \in BPC^2(\mathbb{R})$. Via Corollary 8.5, there exists $T > 0$ such that there exists a maximal solution $u_1 : \bar{D}_T \to \mathbb{R}$ and a minimal solution $u_2 : \bar{D}_T \to \mathbb{R}$ to this (B-R-D-C). Are these solutions equal, or equivalently, do u_1 and u_2 satisfy

$$u_1(x, t) = u_2(x, t); \quad \forall(x, t) \in \bar{D}_T \, ?$$

References

[1] U. G Abdullaev, "Local structure of solutions of the Dirichlet problem for N-dimensional reaction-diffusion equations in bounded domains." *Advances in Differential Equations*, **4**, (1999), 197–224

[2] U. G. Abdullaev, "Reaction-diffusion in a closed domain formed by irregular curves." *Journal of Mathematical Analysis and Applications*, **246**, (2000), 480–492

[3] U. G. Abdullaev, "Reaction-diffusion in irregular domains." *Journal of Differential Equations*, **164**, (2000), 231–254

[4] U. G. Abdullaev and J. R. King, "Interface development and local solutions to reaction-diffusion equations." *SIAM J. Math. Anal.*, **32**, 2, (2000), 235–260

[5] J. Aguirre and M. Escobedo, "A Cauchy problem for $u_t - \triangle u = u^p$ with $0 < p < 1$. Asymptotic behavior of solutions." *Annales Faculté des Sciences de Toulouse*, **8**, 2, (1986), 175–203

[6] T. M. Apostol, *Mathematical Analysis (2nd Edition)*. Addison-Wesley, 1974, London

[7] R. Aris, *The Mathematical Theory of Diffusion and Reaction in Permeable Catalysts: The Theory of the Steady State, vol. 1*. Oxford University Press, 1975, Oxford

[8] R. Aris, *The Mathematical Theory of Diffusion and Reaction in Permeable Catalysts: Questions of Uniqueness, Stability and Transient Behavior, vol. 2*. Oxford University Press, 1975, Oxford

[9] C. Bandle, "A note on optimal domains in a reaction-diffusion problem." *Zeitschrift für Analysis und ihre Anwendungen*, **4**, 3, (1985), 207–213

[10] C. Bandle and M. A. Pozio, "Nonlinear parabolic equations with sinks and sources." Nonlinear diffusion equations and their states, (1986), preprint

[11] C. Bandle and M. A. Pozio, "On a class of nonlinear Neumann problems." *Annali di Matematica Pura ed Applicata* (IV), **CLVII**, (1990), 161–182

[12] C. Bandle, M. A. Pozio and A. Tesei, "Existence and uniqueness of solutions of nonlinear Neumann problems." *Mathematische Zeitschrift*, **199**, (1988), 257–278

[13] C. Bandle and I. Stakgold, "The formation of the dead core in parabolic reaction-diffusion problems." *Transactions of the American Mathematical Society*, **286**, 1, (1984), 275–293

[14] J. W. Bebernes and K. Schmitt, "On the existence of maximal and minimal solutions for parabolic partial differential equations." *Proc. Amer. Math. Soc.*, **73**, 2, (1979), 211–218

[15] W. Bodanko, "Sur le problème de Cauchy et les problèmes de Fourier pour les équations paraboliques dans un domaine non borné." *Ann. Polon. Math.*, **18**, (1966), 79–94

[16] C. Carathéodory, *Vorlesungen Über Reelle Funktionen (2nd Edition).* Chelsea, 1948, New York

[17] E. A. Coddington and N. Levinson, *Theory of Ordinary Differential Equations.* McGraw-Hill, 1955, London

[18] B. Davies, *Integral Transforms and Their Applications.* Springer, 2002, New York

[19] D. A. Edwards, "A spatially nonlocal model for polymer-penetrant diffusion." *J. Appl. Math. Phys. (ZAMP)*, **52**, (2001), 254–288

[20] P. C. Fife, *Mathematical Aspects of Reacting and Diffusing Systems.* Springer Verlag, 1979, New York

[21] A. Friedman, *Partial Differential Equations of Parabolic Type.* Prentice-Hall, 1964, New Jersey

[22] P. Gray, J. H. Merkin, D. J. Needham and S. K. Scott, "The development of travelling waves in a simple isothermal chemical system III - Cubic and mixed autocatalysis." *Proc. R. Soc. Lond. A*, **430**, (1990), 509–524

[23] P. Gray and S. K. Scott, *Chemical Oscillations and Instabilities: Non-linear Chemical Kinetics.* Oxford University Press, 1990, New York

[24] T. H. Gronwall, "Note on the derivatives with respect to a parameter of the solutions of a system of differential equations." *Ann. of Math.*, **20**, 4, (1919), 292–296

[25] R. E. Grundy and L. A. Peletier, "Short time behaviour of a singular solution to the heat equation with absorption." *Proc. R. Soc. Edinburgh Sect. A*, **107**, 3–4, (1987), 271–288

[26] D. Henry, *Geometric Theory of Semilinear Parabolic Equations.* Springer-Verlag, 1981, Berlin

[27] M. A. Herrero and J. J. L. Velázquez, "Approaching an extinction point in one-dimensional semilinear heat equations with strong absorption." *J. Math. Anal. Appl.*, **170**, 2, (1992), 353–381

[28] A. M. Il'in, A. S. Kalashnikov and O. A. Oleinik, "Second-order linear equations of parabolic type." (Russian.) *Uspehi Mat. Nauk*, **17**, 3, (1962), 3–146

[29] A. L. Kay, D. J. Needham and J. A. Leach, "Travelling waves for a coupled singular reaction-diffusion system arising from a model of fractional order autocatalysis with decay. I. Permanent form travelling waves." *Nonlinearity*, **16**, (2003), 735–770

[30] W. O. Kermack and A. G. McKendrick, "A contribution to the mathematical theory of epidemics." *Proc. R. Soc. Lond. A*, **115**, (1927), 700–721

[31] W. O. Kermack and A. G. McKendrick, "Contributions to the mathematical theory of epidemics. II. The problem of endemicity." *Proc. R. Soc. Lond. A*, **138**, (1932), 55–83

[32] W. O. Kermack and A. G. McKendrick, "Contributions to the mathematical theory of epidemics. III. Further studies of the problem of endemicity." *Proc. R. Soc. Lond. A*, **141**, (1933), 94–122

[33] A. C. King and D. J. Needham, "On a singular initial-boundary value problem for a reaction-diffusion equation arising from a simple model of isothermal chemical autocatalysis." *Proc. R. Soc. Lond. A*, **437**, 1901, (1992), 657–671

[34] M. Krzyżański, "Sur les solutions des équations du type parabolique déterminées dans une région illimitée." *Bull. Amer. Math. Soc.*, **47**, (1941), 911–915

[35] M. Krzyżański, "Certaines inégalités relatives aux solutions de l'équation parabolique linéaire normale." *Bull. Acad. Polon. Sci. Sér. Sci. Math. Astr. Phys.*, **7**, (1959), 131–135

[36] J. A. Leach and D. J. Needham, *Matched Asymptotic Expansions in Reaction-Diffusion Theory.* Springer, 2004, London

[37] S. A. Levin, "Models of population dispersal." In: S. Busenberg and K. Cooke (eds.), *Differential Equations and Applications to Ecology, Epidemics and Population Problems.* Academic Press, (1981), 1–18, New York

[38] P. Mattila, *Geometry of Sets and Measures in Euclidean Spaces: Fractals and Rectifiability.* Cambridge University Press, 1999, Cambridge

[39] V. Maz'ya and T. Shaposhnikova, *Jaques Hadamard: A Universal Mathematician.* American Mathematical Society, 1998, Rhode Island

[40] P. M. McCabe, J. A. Leach and D. J. Needham, "The evolution of travelling waves in fractional order autocatalysis with decay. I. Permanent form travelling waves." *SIAM J. Appl. Math.*, **59**, 3, (1998), 870–899

[41] P. M. McCabe, J. A. Leach and D. J. Needham, "The evolution of travelling waves in fractional order autocatalysis with decay. II. The initial boundary value problem." *SIAM J. Appl. Math.*, **60**, 5, (2000), 1707–1748

[42] P. M. McCabe, J. A. Leach and D. J. Needham, "On an initial-boundary-value problem for a coupled, singular reaction-diffusion system arising from a model of fractional order chemical autocatalysis with decay." *Q. Jl. Mech. Appl. Math.*, **55**, 4, (2002), 511–560

[43] P. M. McCabe, D. J. Needham and J. A. Leach, "A note on the non-existence of permanent form travelling wave solutions in a class of singular reaction-diffusion problems." *Dynamical Systems*, **17**, 2, (2002), 131–135

[44] J. H. Merkin and D. J. Needham, "The development of travelling waves in a simple isothermal chemical system. II - Cubic autocatalysis with quadratic decay and with linear decay." *Proc. R. Soc. Lond. A*, **430**, (1990), 315–345

[45] J. H. Merkin and D. J. Needham, "The development of travelling waves in a simple isothermal chemical system. IV - Quadratic autocatalysis with quadratic decay." *Proc. R. Soc. Lond. A*, **434**, (1991), 531–554

[46] J. H. Merkin, D. J. Needham and S. K. Scott, "The development of travelling waves in a simple isothermal chemical system. I - Quadratic autocatalysis with linear decay." *Proc. R. Soc. Lond. A*, **424**, (1989), 187–209

[47] J. H. Merkin, D. J. Needham and S. K. Scott, "Coupled reaction-diffusion waves in an isothermal autocatalytic chemical system." *IMA. Jl. Appl. Math.*, **50**, (1993), 43–76

[48] J. C. Meyer, *Existence and Uniqueness of Solutions to Specific Reaction-Diffusion Problems*, MSci Thesis. 2009, University of Birmingham

[49] J. C. Meyer and D. J. Needham, "Extended weak maximum principles for parabolic partial differential inequalities on unbounded domains." *Proc. R. Soc. Lond. A*, **470**, 2167, (2014)

[50] J. C. Meyer and D. J. Needham, "Aspects of Hadamard well-posedness for semi-linear parabolic partial differential equations." Submitted

[51] J. C. Meyer and D. J. Needham, "Well-Posedness of a semi-linear parabolic Cauchy problem with a non-Lipschitz nonlinearity." *Proc. R. Soc. Lond. A*, **471**, 2175, (2015)

[52] J. D. Murray, *Mathematical Biology (2nd Edition)*. Springer-Verlag, 1993, Berlin

[53] M. Nagumo, "On principally linear elliptic differential equations of the second order." *Osaka Math. J.*, **6**, (1954), 207–229

[54] D. J. Needham, "On the global existence and uniqueness of solutions to a singular semilinear parabolic equation arising from the study of autocatalytic chemical kinetics." *J. Appl. Math. Phys. (ZAMP)*, **43**, (1992), 471–480

[55] D. J. Needham, *Lecture Notes on Reaction-Diffusion Theory*. 2005, University of Reading

[56] D. J. Needham, *An Introduction to Reaction-Diffusion Theory*. Cambridge University Press. To appear

[57] D. J. Needham and J. H. Merkin, "The development of travelling waves in a simple isothermal chemical system with general orders of autocatalysis and decay." *Phil. Trans. R. Soc. Lond. A*, **337**, (1991), 261–274

[58] L. Nirenberg, "A strong maximum principle for parabolic equations." *Comm. Pure Appl. Math.*, **6**, (1953), 167–177

[59] J. Ockendon, S. Howison, A. Lacey and A. Movchan, *Applied Partial Differential Equations (Revised Edition)*. Oxford University Press, 2003, Oxford

[60] A. D. Pablo and J. L. Vazquez, "Travelling waves and finite propagation in a reaction-diffusion equation." *Journal of Differential Equations*, **93**, 1, (1991), 19–61

[61] C. V. Pao, "Successive approximations of some nonlinear initial-boundary value problems." *SIAM J. Math. Anal.*, **5**, 1, (1974), 91–102

[62] A. Pazy, *Semigroups of Linear Operators and Applications to Partial Differential Equations*. Springer-Verlag, 1983, New York

[63] M. Picone, "Maggiorazione degli integrali delle equazioni totalmente paraboliche alle derivate parziali del secondo ordine." *Ann. Mat. Pura Appl.*, **7**, 1, (1929), 145–192

[64] M. H. Protter and H. F. Weinberger, *Maximum Principles in Differential Equations*. Springer-Verlag, 1984, New York

[65] F. Rothe, *Global Solutions of Reaction-Diffusion Equations*. Springer Verlag, 1984, New York

[66] W. Rudin, *Principles of Mathematical Analysis (3rd Edition)*. McGraw-Hill, 1976, Singapore

[67] A. A. Samarskii, V. A. Galaktionov, S. P. Kurdyumov and A. P. Mikhailov, *Blow-up in Quasilinear Parabolic Equations*. Walter de Gruyter and Co., 1995, Berlin

[68] J. Schauder, "Der Fixpunktsatz in Funktionalraümen." *Studia Math.*, **2**, (1930), 171–180

[69] J. Schauder, "Über lineare elliptische Differentialgleichungen zweiter Ordnung." *Math. Zeit.*, **38**, (1934), 257–282

[70] J. Smoller, *Shock Waves and Reaction-Diffusion Equations*. Springer Verlag, 1983, New York

[71] I. Stakgold, "Localization and extinction in reaction-diffusion." *Free Boundary Problems: Theory and Application*, **1**, (1987), 209–221

[72] A. I. Volpert, V. A. Volpert and V. A. Volpert, *Travelling Wave Solutions of Parabolic Systems*. AMS, 2000, USA

[73] K. Weierstrass, "Über continuirliche Functionen eines reellen Arguments, die für keinen Werth des letzteren einen bestimmten Differentialquotienten." *Math. Werke*, **2**, (1895), 71–74

[74] A. J. Weir, *Lebesgue Integration and Measure*. Cambridge University Press, 1973, Cambridge

[75] X.-Q. Zhao, "Spatial dynamics of some evolution systems in biology." In Y. Du, H. Ishii and W.-Y. Lin (eds.) *Recent Progress on Reaction-Diffusion Systems and Viscosity Solutions*. World Scientific Publishing Co., (2009), 332–363, Singapore

Index

Printed in the United States
by Baker & Taylor Publisher Services